The Author

Andrew Caine

Andrew graduated with honours in Marine Biology from University of North Wales, Bangor.

He has been extensively published in various marine-related magazines over a period of 10 years. During this time, he worked in many areas within the marine industry, concentrating on tropical marine biology.

Andrew has now settled down to concentrate on his family and is teaching and inspiring the next generation of scientists.

You may always contact him by email here:

andrewmcaine@gmail.com

Dedications

This book is dedicated firstly to my fantastic family and Tracey without whose unquestionable support this book would not have been possible.

Also, to every marine scientist whose work through time has led us to an understanding of the physical workings of the complex environments within the oceans. Also, to all those who studied the life that exists here and allowed us not only to be amazed at the wonderful animals but be constantly bowled over by the diverse array of new discoveries made every year.

Finally, I would like to thank the animals living in the sea for they are truly amazing.

Picture credits

To obtain the correct illustrations proved to be a search in itself. All I could find were scientific diagrams with the associated scientific terminology, something that has to be avoided in the case, after all this book is for the non-biologist. Then I came across some amazing drawings made by amateur naturalists many over 100 years ago, they are pencil drawings without labels. Yet they are scientifically accurate in their presentation. These are perfect as they give a visual representation of the animals we will be looking at.

It is wonderful to imagine a lady or gentleman of that period sitting down with pencil creating illustrations for the scientific community of that era. They could never of imagined that their outstanding work would help people of a future age, understand the wonderful animals that exist within our oceans. A shared love of the natural world extending through time.

All diagrams apart form two, are from the collection made public by the company Pearson Scott Foresman.

Squid diagram on page 42 is reproduced through the courtesy of Biologycorner.com (Creative Commons Attribution 3.0).

Hydrothermal vent diagram page 112, courtesy of Alexgiovi, CC BY-SA 4.0 https://creativecommons.org/licenses/by-sa/4.0>, via Wikimedia Commons

First edition published 2015

ISBN: 9781520606439

Published by Andrew Caine

Marine Biology for the Non-Biologist

Contents

The Basis of Life in the Oceans

If we look at the earth from a satellite photograph, we can see that about 71% of the earth's surface is covered in seawater. This vast mass of water has an average depth of 3.4 kilometres, the deepest area being just over 11 kilometres deep. If we removed the seas and levelled the solid rock into a perfect sphere, then replaced the water, we would have a body of water encompassing the earth 2.5 kilometres deep.

There is a lot of sea out there, but it is not evenly distributed in area or depth, and neither has it been so throughout ecological time. The landmasses have shifted via plate tectonics so that today, the northern hemisphere has an oceanic coverage of 61% and the southern hemisphere 80%. The continental shelf area surrounding the land constitutes 15% of the sea-covered area, yet over 90% of all life exists in this narrow belt commonly known as the shallow seas. All the rest is the deep sea. The depth of water has been far from stable, 12,000 years ago, the sea level was 125 metres below that of today. By the end of the last great ice age 10,000 years ago, the level had risen to the present day conditions. 10,000 years to adapt and colonise a new habitat sounds like a long time, but in reality it's just a few seconds in ecological time.

But what's the process that allows oceanic life to exist? It's the availability of food that can be passed up the food web to the larger animals. Small animals eat plants, and incorporate organic molecules into their own tissues, and are then eaten themselves. The food webs of land animals are all based on plants taking energy from the sun and converting inorganic molecules to organics. These are then eaten by herbivores, and off we go up the food web. It's the same in the sea, but the plants are different; there's only one type of truly marine rooted plant, the sea grasses. All other production is facilitated via algae, the common sea weeds that we see on the shoreline. However these do not produce nearly enough to keep the seas alive, here we move to the real life-givers, the plankton and microscopic algae.

We can dispel our first old wives' tale: not all plankton are microscopic. Many are, in fact most are, but the fact remains that the word 'plankton' simply means 'drifting life.' This suggests that they may have a form of propulsion, but cannot actively swim against the sea's currents; they are at their mercy. Great migrations occur within 24 hours; they may move up from depths of more than 300 metres to the surface waters and then down again, but always moving with the currents. The largest member is the Lion's Mane jellyfish, which can weigh up to a ton and have a tentacle net of over 400 cubic metres: a large animal by anyone's standards. The one form of plankton that most people will be familiar with is an animal about 6 inches long: the krill, on which the great whales feed.

Let's concentrate on the basis of the food webs, the microscopic algae and animals. There are two separations here: the phytoplankton (algae) and the zooplankton (animals). Zooplankton are again split into two, the holoplankton, which are permanent members of the plankton, and the meroplankton, which are transient members, including eggs and larval stages that can last for a few hours or up to two years.

The algae are the most important members, often single-celled organisms that divide to grow. Like plants on land, they require nutrients in the form of nitrogen compounds, phosphates, carbon (normally CO_2) and sunlight for energy. They can easily be seen in temperate waters around late spring and early summer. Dipping a glass jar into the sea and lifting it to the sky, you will then see hundreds of small spherical shapes. Bingo! Your algae. If you tried this in tropical waters you wouldn't see a thing, because these waters are nutrient deficient and cannot support high densities of algae. What's there is eaten quickly—that's why the water is so clear. Another way of spotting algae is to go out in a little boat on a calm, clear night in mid-June to watch the bow waves. They will glow an iridescent blue, as bioluminescence is produced by alga upon their disturbance. Often beaches are subjected to great quantities of foam being washed ashore, mistakenly taken for pollution. In actuality, it's the remains of millions of dead algal cells: an algal bloom.

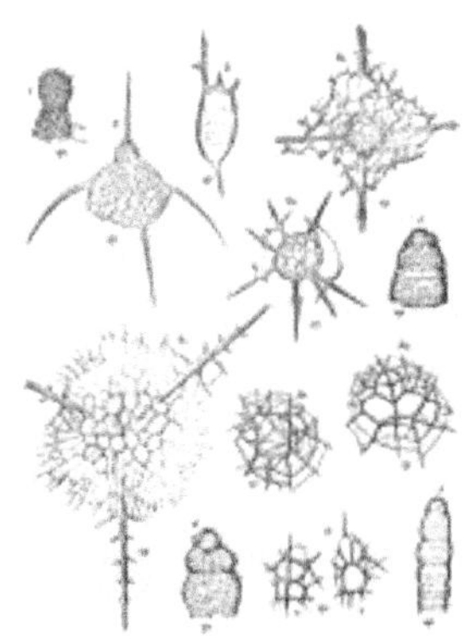

When certain variables, such as the right temperature, nutrient concentration and calm seas all align for a few weeks, the optimum conditions for algal growth are produced, resulting in an amazing division rate of cells: a bloom. Within a week, the concentration of algae can explode from a few hundred per litre to well over 5 million per litre, and it's very destructive for both animal and human health. Some species are red in colour, and blooms of these turn the sea red. Hence, the term 'red tide.' The sheer density of cells and the mucus they produce clogs fishes' and larvae's mouth parts, as well as respiratory structures such as gills, causing mass mortality. Often the first indication of a bloom is hundreds of tons of floating dead fish.

When the bloom dies, all the dead cells sink to the bottom. This can smother all benthic life (life at the bottom of the sea) and cause anoxic conditions, resulting in the death of all benthic animals. I mentioned that algae can be a hazard to human health. Well, that was a bit of an understatement, as it has caused many mortalities, permanent paralysis and other disorders. A few decades ago in America, an outbreak of paralytic shellfish poisoning occurred, resulting in many deaths, and scientists were called in to find the cause. For sixteen years, PhD students went down to the shore every day and collected a few mussels to mash up and look for a toxin, with no positive results. A red tide occurred, and the next day, the mussels were full of a very potent toxin, but the alga did not produce it. It turned out to be bacteria living within the cells. Mussels consumed the rich food source—they can easily eat over 50 million cells per hour—thus storing the toxin and concentrating it to deadly amounts.

The scientists were very pleased with themselves, and justly so, until a local Native American explained that they had known for centuries not to eat the shellfish when the tide turns red! To date, 60 species of algae have been isolated that contain the deadly bacteria, but we'll get to more of that in the marine toxins chapter.

Algae in normal concentrations are the life-givers of the oceans, but when blooming they are the life-takers. Here man has a problem, for the incidence of blooms is increasing both in frequency and duration on a worldwide basis. We don't know if coastal sewage disposal is to blame, enriching the nutrients in the sea, or if the blooms have occurred at this rate through ages of ecological time, with our technological advances in monitoring equipment simply allowing us to see more. Only time will tell.

We now have a condition of a nice algal soup floating and growing in the sea, an abundant food source for herbivorous zooplankton. You will be lucky to see these in your jar of algae as there are less of them and they are often transparent to avoid the carnivorous zooplankton. The herbivores happily swim around, eating the algae and reproducing until one day, BANG! They themselves are gobbled. Here we enter the realm of the first carnivores. Truly, these are sophisticated microscopic monsters, armed with an array of sense organs to detect prey: even the presence of molecules released from the body fluids of an unfortunate herbivore are detected. Their capture mechanisms and mouth parts are truly something out of a horror movie. Coupled with a complex behaviour pattern, these are very successful beasts indeed. Often they lurk deep within the water column during the day to avoid predators, but as dusk approaches, they rise to feed. From underneath, prey is silhouetted against the sea surface and is picked off easily. We have now increased in size from the microscopic algae to comparatively large animals ranging from 0.5 centimetres to shrimp and above.

The smaller zooplankton communities contain both permanent and temporary members, both of great biological importance. The permanent members are a stepping stone to pass organics up to higher levels and keep the herbivore population down, and the temporary members are the eggs and larvae of invertebrates and fish. All the invertebrate groupings have members that produce planktonic larvae in the millions; often only two need to survive to be successful. The belief that larvae are produced in great numbers to ensure that through luck they find the right place at the right time to settle is just not true. An adult habitat is specific; a limpet cannot settle on sand, so larvae have many sense organs to detect the right habitat, though if a prevailing wind blows offshore for a few days at the wrong time, the entire larvae population for that year will be lost out at sea.

How the currents can disperse a species via larvae is best shown by the Australian barnacle. Before the Second World War, the UK waters were free of this exotic species (an exotic species is any which is not native to the country where it's present). With the onset of war, a few ships from Australia were kept in docks and not put out to sea. On the hulls were a population of barnacles. They liked the UK's waters and reproduced there, setting free planktonic larvae. They found a few rocks nearby and settled. They were stronger than the native species and displaced them. They grew and reproduced, and with the current's drift, the next generation moved up the shoreline. They are now common on every suitable rocky shore of the UK after less than 50 years. As you can see, having planktonic larvae often ensures the dispersal of a species.

The crux of the matter is this: when you're in the local fish market and you see a 4 kilo fish, take time to think about the number of algal cells it's taken to produce such a magnificent animal. Ultimately, that's where the energy to produce such a fish has come from.

Soft-Bodied Animals - The Cnidarians

These are the most ancient of the animals that we will discuss, with members belonging to the jellyfish, anemones, hydroids and corals. There are over 10,000 species, living anywhere from the abyssal depths to rock pools, but most abundant in the shallow tropical waters.

Cnidarian body form is very simple, with 2 layers of cells making up the outer body and inner lining. An inorganic layer separates the two. Feeding, for most of them, is via prey capture, facilitated by an array of tentacles, armed with stinging cells known as nematocysts. These cells are kept under hydrostatic pressure, equivalent to diving at a depth of 1400 metres. We still don't know how this is maintained. When triggered, an explosive reaction forces a serrated harpoon into the prey, then a toxin is passed to the prey and its death through paralysis is swift. Many can cause human fatalities, with the most potent toxin 75% more deadly than cobra venom.

The Jellyfish

These are exclusively marine, ranging from the microscopic to a bell diameter of more than 2 metres. Over 200 species are found in all seas, though they are very common in coastal waters. Many species produce extremely painful stings. The Australian box jellyfish produces one of the most toxic substances known to man. Though potent, each sting only delivers a very small amount of toxin; therefore 6 feet of tentacles will kill an adult. Despite this, they have many predators, including basking sharks and leatherback sea turtles.

Often, they are observed in swarms. The possible reasons for this are many, including breeding and abundance of food. We have yet to pinpoint the true answer. The best observations of these jellyfish are made when they are swarming, rewarding a trip in a small boat with the sight of these beautiful animals in action. Watch closely the action of the bell propelling the animal; often you will see the tentacles seize prey animals, see how they are transported to the mouth and gut. After a high tide, a walk along the beach reveals many of these animals stranded, their fate already sealed.

The central bell is the most obvious feature of the body plan, often gelatinous in nature and ranging from transparent to highly coloured. Radial canals emanate from the centre, forming a communication route. Around the underlying edge are the tentacles, which can commonly trail over 10 metres behind the pulsating bell. The mouth is situated at the centre of the underside of the bell and is surrounded by oral arms that function both to move food to the mouth and to assist in reproduction.

These carnivores prey upon any unfortunate animal that comes into contact with the tentacles, ranging from zooplankton to smaller fish. Often large fish can escape with their powers of locomotion, but they may die later because of the toxins. Some of the larger species of jellyfish are prey-specific, which means they prefer one species of animal to eat. The prey often being smaller jellyfish often results in many jellyfish of one species following a swarm of another.

There is an exception in the upside-down jellyfish. As the name suggests, it swims with its reduced tentacles facing the water surface. It has no stinging cells, and does not feed in the normal way, but it contains millions of algae within its oral arms. These provide all the nutrients it requires; however, I will address this subject more in depth when we look at reef-building corals.

For jellyfish to reproduce, sperm and eggs are passed to the oral arms, where fertilization takes place, and a planula larva is produced. This is basically a hairy ball of cells that is released and settles on a rocky surface. It then undergoes a metamorphosis to produce a feeding polyp (like a small anemone). As it grows, elongation causes segments to become apparent. The segments separate to produce the final larva stage, the ephyra. From one polyp, it's common to produce over 15 ephyra, which grow into the juveniles (one egg results in many young). These polyp stages can be easily seen, hanging from the underside of rock overhangs or cliff faces. Sea slugs often eat them, but the stinging cells are not discharged, and after ingestion they pass whole through the gut, and up into the skin layer of the slug. There, the stinging cells perform the same task, and confer predator resistance to the slug. 'Sea slug' is probably one of the most derogative common names for a marine animal, as their colouration provides the most vivid display of any species in the marine environment.

Jellyfish do cause problems to man, especially in the aquaculture industry. Often the animals get caught up in the cages where fish are kept. They struggle to free themselves and in doing so lose parts of tentacles. These free strands cause lacerations on fish, dropping the market price. When sorting the fish, broken tentacles can be 'flicked' onto the worker stinging them. I can personally vouch that a sting from even a relatively harmless species, under the fingernails is a very painful experience indeed. Handle with care.

The Hydroids

These are found in both fresh and salt water. They are either solitary animals or they live in colonies all joined by a common connection. A colony structure is best described as a very long railway line with many stations along it. The common connective structure is like the line, and each individual animal is like a station. Extensive rock areas can be covered by the same species, which often resemble a fern. They commonly attain a height of between 5 and 15 centimetres, though much larger and more dangerous species exist. Apart from rocky substrate, they are found on shellfish, crabs and seaweed fronds, and sometimes are free floating. Take a powerful hand lens down to the shore, find a seaweed with brown moss growing on it, place this in a jar of water, and you will be able to observe the structure more closely.

The best-known hydroid is the Portuguese man-of-war, commonly thought to be a jellyfish. In fact, it is a hydroid colony composed of tens of thousands of individual animals. It has no bell, and is supported by a float under which the 'tentacles' trail. Each tentacle is in fact a very long railway line armed with separate feeding animals, each passing some

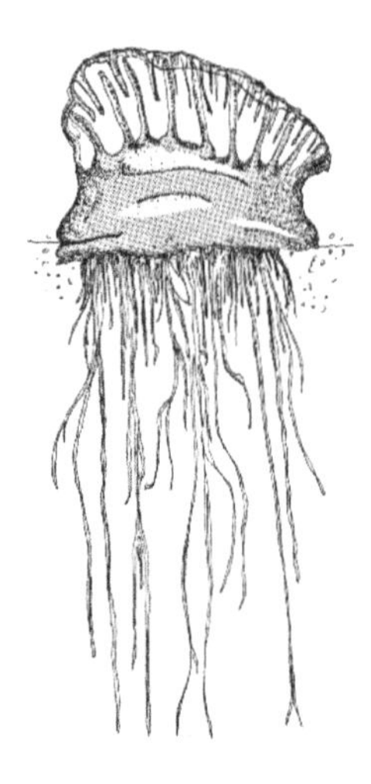

energy up the connective structure to maintain the carbon monoxide-filled float. These colonies are common in the tropics and sub tropics, with species' float lengths ranging from 1 - 30 cm. They can also be seen in the North Atlantic when a storm blows them out of the Gulf Stream. If you encounter a large one when taking a dip, the best advice is to get the hell out of its way, as a sting can be dangerous. Fatalities are mostly due to drowning because of the resulting muscle seizures and cramps, not directly caused by the sting.

Another hydroid commonly mistaken for a different animal, one that has caused suffering to many divers, is the fire coral. As the name suggests, it comes with a very painful, burning sting. Like corals, it builds a calcium carbonate skeleton. The difference is that it's surrounded by living tissues, whereas a coral's calcium is on the outside. A hydroid it is; a coral it is not.

An animal's body plan is made up of three polyps, each with a distinct function: feeding, reproduction and defence. The connective structure is called the stolon. In this colony, two main growth forms are operant. First is monopodial growth, in which single animals are arranged along the length of the stolon. Second is sympodial growth, where a common basal stolon runs along the substrate with fern-like growths budding off, each consisting of an extension of the stolon and many individuals.

The defence polyp is the only structure to contain stinging cells; therefore, it carries out food capture and passes the prey to the feeding polyp for ingestion. Prey consists of mainly zooplankton, but the larger-sized colonies can take small fish.

Hydroids reproduce in two ways, asexual and sexual. The colony increases in size via asexual reproduction, in which buds form on the leading tip of the stolon. These buds form clones of the original polyp, then the stolon increases and another bud forms, and so the colony grows.

Sexual reproduction ensures the mixing of genetic information, thus allowing evolution to occur. The ways hydroids reproduce sexually varies among species and may include the direct shedding of eggs and sperm into the water column. Fertilization then occurs and a planula larva is produced. This larva settles and metamorphoses into a young hydroid to start a colony. The reproductive polyps can also shed fully-formed medusae, which look like microscopic jellyfish. This allows dispersal through water currents. The medusae then release eggs which develop into a planula, and the cycle continues. A seven-centimetre colony of one species was studied that released 4450 medusae in three days. It is a very prolific animal indeed.

The Anemones

These belong to the same family as corals, with the same distinguishing feature: having six or multiples of six tentacles. Anemones are just as widespread and diverse as coral as well. They can be found in all seas and at all depths, but intertidal species are often smaller than the sublittoral (constantly immersed) animals. This is because the intertidal species possess a collar region where the tentacles can be retracted when the tide is out, reducing their risk of drying out. They also have to withstand wave action. They can be seen at the lower regions of the tidal range, often on overhangs of rocks or in rock pools. A blob of jelly is a retracted anemone.

On spring tides (the lowest tides, every 2 weeks), the sublittoral species can often be located under kelp, on rocks, through shallow wading. They are extremely aggressive toward other species, often fighting each other with specialised searching tentacles when they meet. This behaviour can result in the death of one, through stings. Normally the faster-growing species are not as aggressive as the slower growers, preventing the latter from becoming dominant in one area.

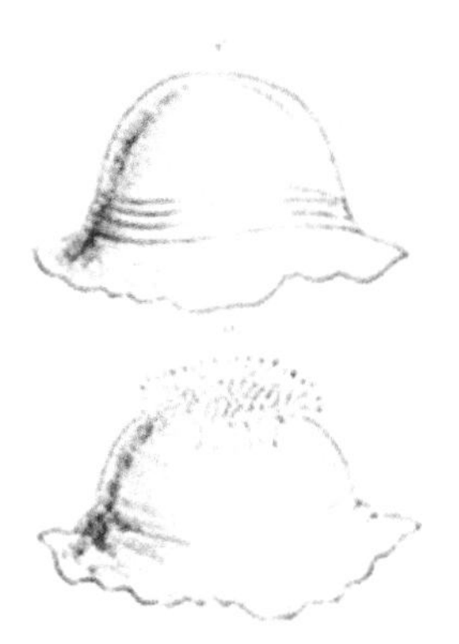

Anemones have a size range from one millimetre to three metres across, and can be found in soft sediments or on any hard substrate. Pier supports are a good place to look for them. Their ability to settle on any surface causes great commercial cost to oil companies and power stations, as anemones constantly settle in and clog any system with sea water flowing through. Over the body a mucus coat is secreted: this is the mechanism that allows clownfish, shrimp and other animals to live within the tentacles. These animals produce the same mucus coat, thus not being recognised as potential prey. Anemones can confer protection to hermit crabs by living on the shells. When the crab changes its shell, it removes the anemones and places them on the new shell. Some species can be very long-living, attaining an age of over 60 years. Not many do, however, as they are often eaten.

The anemone body plan has many variations on the typical polyp form, that is, a thick circular base, a cylindrical body topped with a ring of tentacles, and a central mouth. The pedal disc (the base) is often muscular for attachment and movement, as these are very mobile animals indeed. It is also utilised in asexual reproduction. When located, don't try to remove them, as the basal disc is so strong you will damage the animal if you remove it. The body is often highly contractile, allowing expansion to envelope large prey items, and contraction to expel water and waste through the mouth. The body may be very long and flexible, up to a metre in deep-sea species, or very short in the shallow water species. The tentacles also are very varied, ranging from a thick leathery appearance to feathery structures for filter feeding. The tentacles' length varies from over two metres to less than a millimetre. Here a sticky mucus is produced, which can be tested safely by touching them with your fingers; the skin on our fingers is too thick for the stinging cells to penetrate; however, the soft skin of our underarms is not. Feel just how sticky they are, and watch how the tentacles react to a potential prey item.

Anemones are voracious carnivores, trapping and subduing prey with their tentacles and stinging cells. They are capable of ingesting fish larger than themselves, but for smaller sized items they have a ciliated channel (a depression in the body running to the mouth lined with many stiff hairs called cilia) to help them feed. The cilia all move in a wavelike fashion, transporting the small prey to the mouth. The tentacles are all individually controlled, and when not feeding many may be deflated, whilst others are inflated in the current. On encounter with a prey item, the sting enters the body, and chemicals are released that stimulate the other tentacles, causing a rapid reaction. Almost instantaneously, all the tentacles are inflated and wave around, ensuring no escape for the prey.

Again, some species contain algae that passes on nutrients to the host, but prey capture is also required to provide the animal with essential nutrients that the algae does not produce.

Locomotion; anemones are often thought of as sessile animals (not moving), but nothing could be further from the truth. They can move in a number of ways. Already mentioned is their slow gliding over substrate via the pedal disc. Other methods include crawling on their side or walking upside down on their tentacles. Some can even swim for short periods by use of the tentacles. One oceanic species secretes a substance from its pedal disc that acts as a float, suspending the animal upside down at the surface. These can be found in the middle of the Pacific, thousands of kilometres from any land.

In regard to sexual reproduction, most species are hermaphroditic (able to act as male or female) though they can act as only one sex at a time. Eggs may be fertilized within the body, or shed into the water, depending on the influence of the lunar cycle. A planula larva is produced, which disperses the species. Tentacles only develop when settled; this represents the change from larva into juvenile.

As far as asexual reproduction, there are two common forms, both of which often result in whole rock surfaces or other extensive areas being colonised by clones of the original settler. Longitudinal division occurs, in which the animal splits down the body, resulting in two thinner animals. The second and more common is pedal laceration, in which the parent leaves behind a small piece of the pedal disc when moving. From this small section of foot cells, a new clone arises.

The Corals

In 1723 a French scientist proposed that corals were indeed animals and not underwater plants! How dare he! He was ridiculed by the scientific community of the day, and gave up science.

Corals are probably the most famous of the marine invertebrates, though they are known not for themselves but for the structures they produce, and the array of colourful fish that inhabit their reefs. Making the areas very popular for tourism, it's a shame that the reefs attract such commercial interest, because it will be their downfall. They are grossly understudied, as SCUBA diving has only been developed to a high standard within the last 30 years, and the fact that reefs are most often found in underdeveloped countries, where cash for research not leading to profit is not available. This is another area of public misinterpretation because science knows relatively little about coral reefs. To illustrate this, it has been calculated that species on the world reefs should number 423,000, of which less than 10% have been named, and the majority of these have never been studied. The only intensive study around reefs is for bio-research, which involves taking animals and mashing them up, as many produce chemicals that have the potential to cure cancer and other diseases. On the whole, we know very little.

We have two separations here: the soft corals and true corals, with the distinguishing feature being the development of tentacles. The soft corals, or octocorals, have eight or a multiple of eight tentacles, and the polyps are enclosed in a flexible framework. The tentacles always have branches on the side for filter feeding. The true corals have six or a multiple of six tentacles, and are enclosed with a calcium carbonate skeleton. Two separations occur here between ahermatypic (non-reef builders) and hermatypic (reef builders and stoney corals) corals, which all contain algal cells within their own cells.

The soft corals and non-reef builders have a global distribution at all depths. The soft corals can live in colonies in many different shapes, from elaborate fans to single strands many metres in length. The diameter reaches a maximum size before 30 years, after which breakages occur. They are active during the day whilst the other corals are mainly nocturnal. Non-reef builders take a solitary form, with a single polyp enclosed in a small calcium carbonate structure. The tentacles are often elongated to act as legs, to either move up with the rising sediment to avoid being buried, or to drag the carbonate 'house' along the bottom. Deep-sea species are extremely slow growing, individuals recorded at being over 300 years old.

The hermatypic corals are distributed within the tropics of Cancer and Capricorn, as well as in the Red Sea. The best growth is in clean, clear, well-oxygenated water at a temperature of 23 – 25°C. They do not tolerate high turbidity, temperature above 30°C or below 18°C or low salinity. They are found in surface waters down to 100 metres with most growth between 10 and 30 metres, forming colonies that can weigh over 100 tons. There are three basic colony growth shapes: massive (rounded), branching (treelike) and tabular (flat) with coral species being able to develop any combination of these growth forms. Growth of a colony is achieved by depositing calcium carbonate on the surface, thus increasing the skeleton's size. The dead structures found at depths of 100 metres were thriving in surface waters when the Romans invaded England.

The Stony Corals

We have already discussed that the bases of the marine food chain are phytoplankton and algae, which require inorganic nutrients from the surrounding sea. We have a problem in the tropical waters, then, since there are very few nutrients and phytoplankton, so how do coral reefs exist in waters that cannot support the high density of life found there? This is an extremely important question, because without corals, the oasis of life that we know would not be there. What allows the corals to exist? The answer lies within the coral polyp and the population of algae that lives there, this algae is the basis of the reef ecosystem. Older texts refers to the algae as one species within all corals and other Cnidarians, but research has proven this to be untrue. The algae is collectively referred to as *Symbiodinum sp*. The 'sp' refers to species, as many are unnamed. In fact, only 10 species have been positively identified and named; the rest are just sp. They exist at a density of six million cells per cubic centimetre of polyp tissue, and are also responsible for the process of calcification as well, so their importance cannot be underestimated. Without the presence of the alga, the land which they surround would look completely different today because of coastal erosion. The Maldives and other atolls would not exist.

The algae are found in the cells lining the polyp's gut. During the day, using photosynthesis, they leach simple compounds from their cells into the cells of the polyp. The compounds include glucose, glycerol and amino acids, which are utilised by the polyp, either directly in energy-releasing reactions or as building blocks for more complex molecules such as proteins. However, the algal cells need nutrients themselves. These are supplied as waste compounds from the polyp's own metabolism, so we have an extremely efficient recycling system. The polyp feeds the alga and the alga feeds the polyp. In the surface waters, the algae can supply over 100% of the coral's needs, but with the deeper corals, less light penetrates, reducing the possibility of photosynthesis. Thus the corals become less dependent on this source of nutrition. Even with over 100% of their nutritional needs, the corals must actively capture zooplankton to supply the cells with essential nutrients that are not supplied by the alga. So this symbiotic relationship allows an oasis of life to exist within a nutrient-deficient sea. However, it's not a simple as that. The algae cells divide at 9 times the growth rate of polyp cells, so really the cells should burst. The algal biomass is regulated to a constant level. We don't know how, but it's unlikely that they are consumed as no evidence has been found. We know that corals are capable of releasing all the algae to the surrounding water when stressed. In the Caribbean a few years ago, a temperature rise occurred which resulted in many coral deaths through 'bleaching'. In fact, the corals expelled all their algae and died. The current belief is that old cells, which are not as efficient, are expelled from the body and act as a source of nutrition for others.

The growth of the colony is by asexual reproduction of polyps and an increase in the calcareous skeleton. Each polyp sits in its own little depression called a corallite. The pattern of this cup is used in identifying species, as each is different. Microscopic crystals have been observed leaving the polyp and being deposited here, so we know that growth is occurring. From measuring the growth over 24 hours it was discovered that, at night, growth was reduced by 14 times from the growth at midday, indicating that algae were also responsible as they are more active in the daylight. Here we have to use a bit of chemistry to explain the process. One of the main substrates that the algae utilize is waste CO_2 from the coral. This increases the amount of carbonate within the coral cell fluids to dangerous levels. Through a series of chemical reactions, the coral forms a carbonate crystal, thus reducing the carbonate levels and producing a building block for growth. As in the above situation, this is an ongoing process, with the algae and coral working together. Growth is by no means conservative, but always slow. The branching corals are the fastest, growing at the tips; these can attain 10 centimetres per year maximum. The slower growers, the massive and tabular corals in deeper waters, may only grow one millimetre a year. Asexual budding of the polyps causes the circumference to expand and the old polyps are covered by the new growth; only the surface corallites contain living tissue.

Sexual reproduction of stony coral is the same as described for the anemones, and dispersal of the species is achieved. This is best seen in the colonisation of shipwrecks: within 10 years, small colonies are evident all over the superstructure, and a new mini reef is produced. On the west coast of Africa, offshore oil terminals are now covered in corals and their associated inhabitants. This area is species-deficient, so this new unique production is utilised by local fishermen as a source of protein.

Shellfish - The Molluscs

These are the shellfish which we all know and love, not only as animals, but to consume. Sea slugs, octopuses, squid and cuttlefish also belong to this group. There are over 50,000 species of molluscs with a fossil record going back over 500 million years. The groupings that we shall examine include the gastropods (limpets, whelks and winkles), bivalves (mussels, clams and cockles) and cephalopods (octopus, squid, cuttlefish and the nudibranchs (sea slugs).

The Gastropods

These are characterized basically by a shell plonked on top of an animal. We have three sub groupings here: the Archaeogastropods (limpets), Mesogastropods (winkles) and the Neogastropods (whelks). However, with the differences that exist between the groups, there are evident adaptations on a common theme.

The gastropods have one basic problem with locomotion when you think of the environment in which they exist. This is that there is always a force (current) acting on the animal. If dislodged, they would be swept off the substrate, and when that occurs there is always a waiting predator who would relish a snack. So how do these beasts not only hold on to a surface but actually move with only one foot?

The answer lies in a special mucus that is produced and secreted over the base of the highly muscular foot. We have all seen snail trails on land—well, it's the same in the sea. The mucus is composed of 94% water and 4% protein. The secret is in the molecular makeup of the protein and the bonds in the molecule that hold the protein together. These bonds are broken when stress is applied, and the mucus then acts as a lubricant, allowing sliding movement. When the stress is released, the molecular bonds reform and the mucus acts as a glue that holds the foot to the substrate. So we have a highly muscular foot coved in a sticky mucus. A small muscular wave then appears at one end of the foot and progresses down the length, applying a stress to the molecules as it progresses. This allows the mucus to behave in a lubricant manner along the wave and as a glue when the wave passes. With many waves occurring at any one time, parts of the foot are moving whilst other parts are holding on. In this manner, locomotion and stability are achieved at the same time. So there you have it: an animal that is able to hold on to a surface and move along it with only one foot, truly an amazing biological feat.

Every method of feeding is displayed by these animals, from active predators (some catch and eat fish) to deposit feeding and scavenging. However, there is one common characteristic in the gastropods: all food is obtained via a specialized structure known as the radula. The basic radula is like a long strip of sand paper which is placed on the surface. Food is then rasped off and passed to the mouth. The number of teeth present on the working part range from one to 750, and like shark teeth, they are replaced when worn.

The Mesogastropods

Winkles and topshells are mainly herbivorous and exist on a worldwide scale often in very high densities in the intertidal zone and below. They utilize both macro and micro algae depending on the species; one species eats a specific food, so many species are able to coexist without competing. Macro algae (the large seaweeds) often have grooves cut out of them where grazing has occurred and the radula teeth have gouged out a slice. Over rocks in the intertidal zone is a microscopic algal film which we cannot see, but that is actively grazed.

The Archeogastropods

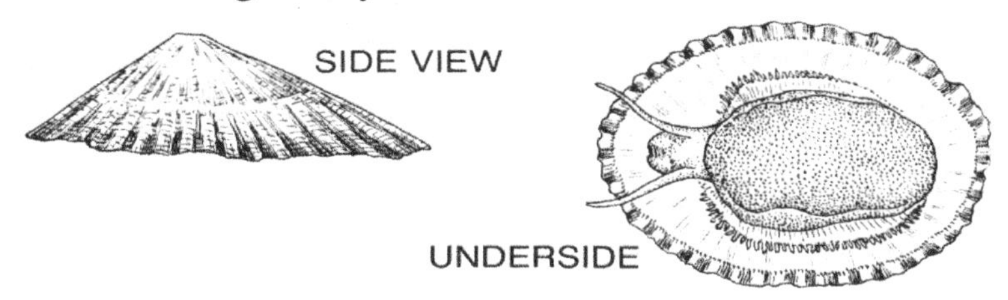

Limpets are the oldest members of the shellfish alive to date, with one species dating back 260 million years, again found on hydrothermal vents. There are two forms, the homing and non-homing species. However, all intertidal species feed nocturnally at low tide. This behaviour avoids predators from both land and sea. The non-homing species graze over the rock surface and settle down anywhere in daylight or when covered by the tide. The homing limpets have a home scar, which is a depression in the rock surface that fits their shell outline perfectly, and increases with growth. The beasts have a territory around the home scar, the size of which depending on the species ranging from 10 to 150 centimetres. Some species have been shown to actively fight for the rights to a feeding territory, pushing the loser off the rock to make a snack for a passing crab. These home scars can be seen by placing a blunt strong knife under the shell and removing the limpet, but always replace the animal once you have seen it.

The Neogastropods

Whelks and cone shells are the predators of the group, with whelks taking a fancy to bivalves and barnacles and the cone shells eating more mobile prey such as worms and fish. The radulae have become specialized and contained in a moveable syphon. If you turn one over, you will see that the circular opening of the shell has a siphonic canal, a characteristic of these carnivorous shellfish. Whelks have a boring radula and a sulphuric acid-secreting gland. These work together to penetrate the shell of an unfortunate victim. The drilling is achieved by secreting acid for around 30 minutes to soften the shell, then the radula drills for a few seconds to remove the weakened part, then the process is repeated. It can take up to eight hours for a 2 millimetre shell to be penetrated, so these animals work for their food. Finally, the syphon is pushed into the shell and the soft flesh is ripped apart and ingested. The shape of the hole is always bevelled, so you can easily see where these creatures have been feeding on the shore.

The cone shells, which are mainly sub-tropical to tropical in distribution take a more mobile prey, but how does a slow snail catch food that is always faster and often bigger than it is? Either through ambush or deceit. For example, a cone shell that lies buried until it senses a fish, then exposing its brightly-coloured syphon and waving it about. The fish thinks it's food but it's in for one hell of a shock. The radula here has one hollow, loose tooth; spares are kept in an internal sac. A tooth is passed to the end of the syphon, and when a fish comes to nibble it is harpooned, often in the mouth. A very potent toxin is forced through the hollow tooth, which acts like a hypodermic needle, and the fish is dead within seconds.

Human fatalities have occurred when handling these and death is certain within 4 hours if not treated.

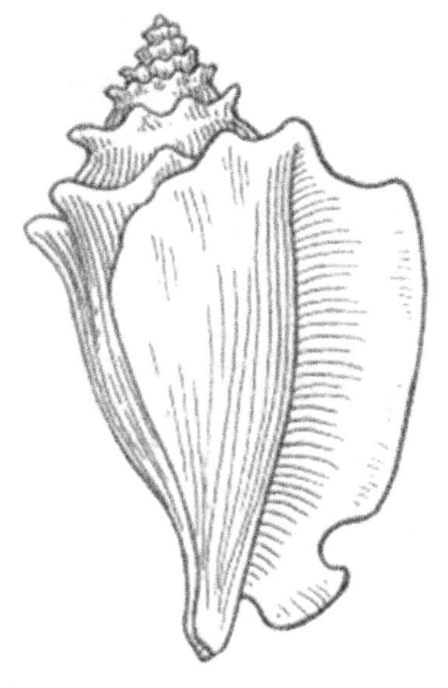

Archeogastropods show by their reproduction that they are the oldest molluscs, as they are broadcast spawners, with fertilization in the water column and planktonic development. However, as we move up the evolutionary tree, internal fertilization is the rule. Mesogastropods lay bundles of eggs that hatch out into either feeding or non-feeding planktonic larval stages. The non-feeders remain in the plankton for only a few hours, but the feeders may remain for up to a couple of months. Here the shell is modified and acts as a sail helping with buoyancy and dispersal of the species. Neogastropods are the most advanced, having done away with all larval stages, giving them advantages including lower mortality—less energy spent on producing eggs, and when the young hatch they are already in the adult habitat where plenty of food is available. Eggs are laid in cases, usually 10 to 80 laid in one season with 4 to 10 eggs per case. On hatching, a fully formed baby beast emerges. The cases are often yellow in colour and can be seen under rocks or rocky overhangs in the intertidal zone.

The Nudibranchs (Sea Slugs)

These are shelless gastropods and are probably the most beautiful marine animals, their vivid colouration and cerate (tentacle-like upper surface) make these a wonder to find. The cerate can be blue, red, green, yellow—really any colour—and the shape varies from thick and stout to long and feathery. All these combine to produce a visual wonder of the oceans, the 'wonder' part is not only restricted to the visual, as the word certainly applies to nudibranchs' feeding. The vivid colouration acts a warning to others not to eat them (they are slow-moving and have no shell for protection). Their protection comes in the form of glands running along the body secreting sulphuric acid and other nasty substances. They also can contain nematocysts to sting any beast trying to take a nip.

Nudibranchs are mainly active carnivores, eating sessile prey such as hydroids, corals, jellyfish polyps and sponges. Some are herbivorous; they are normally the colour of the plant they eat. The nudibranch, though a gastropod, is equipped with a set of jaws to nibble away at its food source, which is always species specific (one species of nudibranch only eats one species of prey).

Here we come to the 'wonder' that was mentioned above, and may be the beginning of a new symbiotic relationship. When a slug eats a polyp containing stinging cells, somehow they are not discharged or digested when in the gut. When you remember that these are normally discharged on touch, this is unbelievable; yet somehow it happens. When in the gut they are not excreted but retained and passed through the body up into the cerate, where they provide a defence for the animal. Even immature cells are translocated, later maturing into a fully-formed stinging cell. We haven't finished here yet; the herbivorous nudibranchs also show a unique behaviour. In a cell there many small parts acting together to make the cell function, like organs in our body. The chloroplast is the part of the plant cell which traps sunlight and releases energy. When the algae is ingested instead of digested, the chloroplast then becomes incorporated into the gut lining and functions normally for a few weeks, supplying the animal with energy. After a few weeks the chloroplast starts to break down and is then digested and replaced. How the nematocysts and chloroplast are translocated and allowed to function is a mystery.

The Bivalves

This grouping of shellfish is probably the most well known, including mussels, clams and scallops, to name a few. They provide a tasty morsel for many people, whether in an expensive restaurant or as a bite on holiday at seaside resorts. Their popularity is increasing and now provides a vast aquaculture industry worldwide.

The soft body is encased by a single shell which has been split into two sections called valves, hence the term bivalve. Each valve is hinged by an abductor muscle used for opening and closing of the valves. The splitting of the shell has allowed many to take on a thin appearance, allowing rapid movement through soft sediments. Most species then remain buried for most of their lives, so unless you dig for them you won't see them. Oh what wonders we walk over on a beach! However, a few species are adapted for a more visible existence, including mussels, clams and scallops.

The soft bottom species are well adapted to living in the often shifting environment. The most familiar bivalves in the sediment are the common cockle and the razor shell, and here we have two good examples of different strategies. The cockle has a fat, ribbed shell and as such is a shallow burier, using its foot to move through the sediment. This also affords predation escape, since crabs and birds like them on their menu, the animal is able to push down on its foot and jump away. The razor shell shows great adaptation for life here; it has become elongated, thus the body is thinner, allowing a more streamlined shape. The foot here allows a very rapid burying speed to escape becoming a meal. Many species live in soft sediments with different species occupying different tidal heights. Take a spade and a garden sieve to the shore, dig down 10cm, and sieve in water to allow the sand to fall away, revealing a range of unseen life. Try this at different tidal heights to see the zonation of animals along the shore.

There are many species living on hard surfaces. These include the mussels, clams and oysters. They attach in one of two ways. First, they attach with one valve to the surface and the bottom valve often enlarged, increasing the surface area for attachment, with the upper valve effectively being a lid. The other way is through the production of byssus threads, which are extremely strong and best seen in mussels. These are also used in larva dispersal, where a single thread up to 30 cm is secreted. This acts as a drag float, giving the baby a greater velocity in the water and reducing sinking, allowing greater dispersal. This behaviour is also shown in the dispersal of spiders, where babies leaving their eggs climb upwards and produce a long thread of web. This catches the wind, and off they fly on their travels. Offshore oil rigs are often covered in mussels hundreds of kilometres offshore from any bivalve population.

There are some unattached species that inhabit hard and soft substrata. These have left the sessile lifestyle and represent the most advanced bi-valves, the most familiar being scallops. These have the ability to be quite nasty to us humans when disturbed. If you poke an open shell, the valves snap shut and they spit at you, so you end up with a wet face and sore finger. However, if we look at the biological implications of this behaviour, we can see how effective it is. The snapping shut of the valves has two functions: it protects the animal from a predator, and it forces water out, propelling the animal up to one metre away and out of danger. This swimming ability and unattached lifestyle allows escape—without this ability, the bivalve would have been doomed.

There are also quite a number of species that live within solid substrates such as wood, sandstone and coral structures. These are known as the boring bi-valves, but boring, they are not. When a larva settles, it starts to burrow. This represents a most important stage of its life. When inside a burrow, predation is a very low risk, so it really has to move to avoid being eaten. Once a burrow has been dug, the beast is trapped, because as it grows the burrow increases in size, but the entrance doesn't expand to allow the animal back through. Therefore, the animal is always a lot bigger than the entrance. The rear of the shell is used as the cutting tool and is often serrated for this, as the burrow is enlarged and lengthened. The sides are coved in a shell secretion to form a smooth lining. They filter feed with long syphons reaching the burrow opening and sucking in food, the animal remaining completely safe inside the rock.

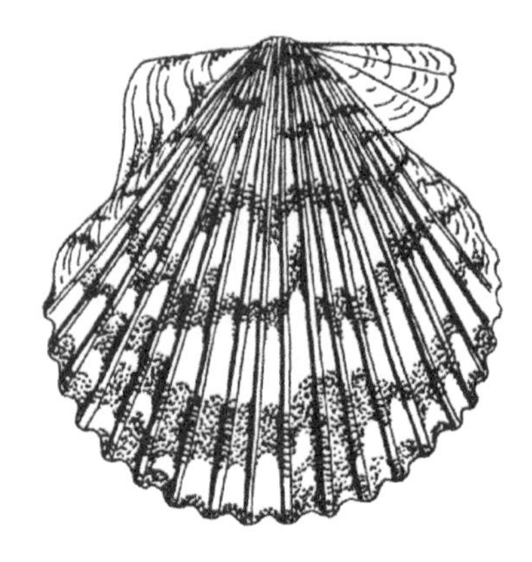

The radula has completely disappeared in all species of bivalve, so it has had to adapt in order to feed. We shall work up the evolutionary tree, beginning with deposit feeding. There are two main types here. The first is where mucus-covered tentacles are pushed into the sediment that organics stick to. These are then transported by stiff hairs known as cilia to the labial palps. These sort the food for ingestion. The other method has been developed by the deep buriers. In this method, a syphon is pushed up to the sediment surface, where it literally hovers over the sediment surface, transporting the food back to the deeply buried beast. These remain safely buried, happily feeding until a fish nips off the syphon, which is then regrown. Many of these species are known as gapers due to the fact that when the syphons are retracted, the body is too big for the shell and so gives the appearance of a gaping animal. Filter feeding is the main approach used by bi-valves occupying hard surfaces. Here the gills have become greatly enlarged and occupy most of the shell. They facilitate gaseous exchange in the uptake of oxygen and expulsion of carbon dioxide, and the capture of food for ingestion. The gills are coved in cilia of different sizes. Some are used to block sand grains and other particles that would damage the delicate gill, and others sort the food and pass it to a ciliated food channel, which is then used for ingestion. Water is brought in via an inhalant syphon and extruded by a smaller exhalant syphon. The smaller size allows the leaving water to leave at an increased velocity (it's like putting your finger over the end of a hose) reducing the risk of re-ingesting used water. The syphons can easily be seen in a tank taken to the shore; mussels are capable of ingesting over 50 million algal cells per hour, so the cilia over the gills must work quite fast!

Some species are carnivorous. These mainly inhabit the deep sea, where food is not found in great supply. Here, a small animal is detected on the sea floor. The inhalant syphon shoots out from the sediment and sucks up an unfortunate. Another design is where the syphon forms a hood which envelops the prey and drags it down for ingestion.

Reproduction; most bi-valves show separate sexes and external fertilization as the rule. However, a few species are hermaphroditic—mainly the boring species, as they cannot actively find a mate of the opposite sex. Some develop the ability to swap sexes; depending on the requirement at a particular time, a male may turn into a female and then back again many times in one year.

The larva all have open shells present that act as floats and sails in the water, aiding buoyancy and dispersal.

The Cephalopods

With over 600 species, these are commonly known as the shelless molluscs; however, this is not completely true, as the most primitive have an external shell. One stage up the evolutionary tree, and the shell is reduced in size and density and is located internally. The most advanced have completely lost the shell. The average size range is between 5 and 65 centimetres, but if we disregard jellyfish tentacles, the longest invertebrates belong to this group.

The Nautilus

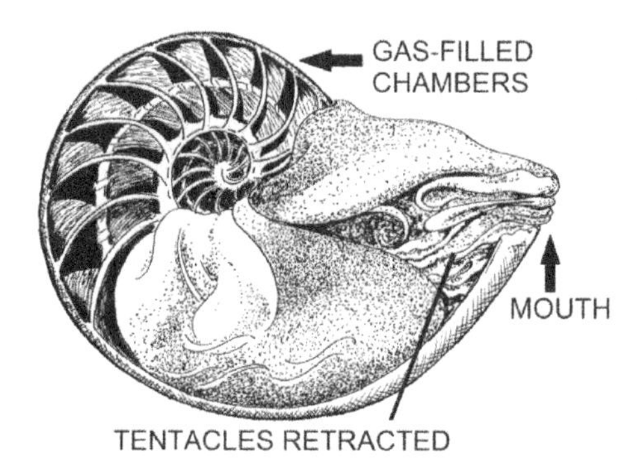

These are the most primitive of the cephalopods alive, with a fossil record going back over 250 million years and 4 species persisting today. It is a relatively clumsy animal with limited powers of locomotion and 32 tentacles for feeding and mating. This animal is probably near the end of its species' lifetime, as it is a very rare and ancient beast found only in a few places in the tropical Indo-Pacific.

The two other remaining groups are very successful indeed: the squid/cuttlefish and the octopods, commonly known as octopi. They have many similarities so they shall be discussed together.

The Squids, Cuttlefish and Octopi

The largest known invertebrates are present here, often known only to exist because of their remains washing ashore or found in the guts of other animals, but never being observed alive. Some species have been observed but never caught for study, as they are just too fast. So we have the exciting situation that many species must exist that remain unknown and are most likely the source of many old tales of sea monsters. Given the size, if a person ever came face to face with a giant squid, believe you me; it would be a monster to anyone.

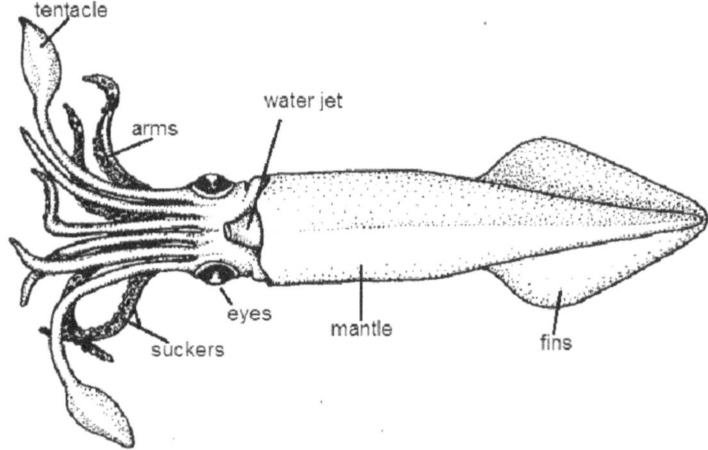

The giant squid was thought to exist by enormous beaks found in the stomachs of sperm whales in the whaling days of the 19th century. This was responsible for the re-emergence of tales of great monsters taking ships and eating all aboard, as the beaks found might have been from a 'baby' for all they knew. There may be animals of that size existing in the deep, but the giant squids that we know are not one of them. No live beasts of this size has ever been observed, as they would have detected our presence and 'done a runner' long before we could detect them. However, a few dead ones have been washed ashore and studied; one such squid washed ashore on the coast of Norway in 1954 with a total length of 16 metres and a body circumference of nearly 4 metres.

This giant size is not restricted to squids, for following closely are the octopi, the most familiar and studied ones living in the Pacific, and the largest actual recorded one measured at 10 metres. Science recognises this as the largest species, as it's been caught and studied, but in the Japanese Sea lives a bigger beast which has eluded capture to this day. Observations have estimated a total length of up to 15 metres, just one metre behind the giant squid in Norway. Which is the largest, the squid or octopuses? Only time and good fortune will tell. I only hope I never come across one in the wild.

These animals have developed two specialized cell types: light-producing chromatophores and ink glands that are used in social behaviour, reproduction and escape from predators.

The chromatophore is filled with a species of bacteria that exists in a symbiotic relationship with the animal and is the source of this cold light. Unlike a light bulb, this light produces no heat, and the colour is always blue. The light emission is controlled by oxygen supply to the bacteria; thus, the blood capillaries which are connected to the chromatophore are switched 'on' and 'off' when required. When blood flows, light is produced. The light passes through a lens to increase the brightness and then through a filter, allowing different displays of colour to be seen. Research is currently being employed to understand these displays with the belief that it reflects the mood of an individual and thus assists reproduction and social behaviour. When approached by a predator, the animal in a split second turns from a dull to a bright colour thus distracting the predator and allowing a quick retreat to safety.

Predator avoidance is also facilitated by the ink gland. This gland produces an alkaline black substance that performs a dual role. On discharge, it first impairs the vision of the hungry beast whilst the animal makes a quick retreat. Then, the alkaline substance acts on chemosensory glands, further blinding the attacker. A gob of ink is not what it expected.

These beasts swim by jet propulsion, which is produced by inhaling water through a large opening, then forcing the water through a much smaller orifice. This produces a jet force, propelling the animal. The exhalant syphon is highly manoeuvrable and along with the aid of fins, it can act as a sort of rudder in directing the water outflow and thus the direction of an individual. Speed of the animal is controlled by the rate at which water comes through. This is normally set at cruising, but when a cephalopod is alarmed, its reaction is instant: one minute, it's there; the next, it's gone. This is not used continuously as it uses a great deal of energy, so squids and cuttlefish supplement this with fins running down the sides of the body whilst octopods crawl over the bottom.

These animals have a distinct feature that's unique among invertebrates: they have developed a highly complex eye. This allows a very high standard of vision in the marine environment and is used to find prey. With this, the development of chromatophores goes hand in hand as they need good vision to see the colour changes and mood of another animal. Prey capture is via tentacles and arms.

Squids have eight short arms covered in suckers and one pair of long tentacles which have suckers only on the flattened ends. These shoot out and attach to prey, which is now doomed. Octopi have only eight arms which are all the same size and covered in suckers, and independently controlled. Once caught, the prey is enveloped in these arms, and a powerful beak soon finishers off the helpless animal. This is why I do not want to ever see a giant squid face to face! To add to the action, poison glands are often present, most of which do not affect man, but which soon finish off any prey. The Indo Pacific blue ringed octopuses can cause an agonising death to humans in under two hours, but you'll read more on that in the toxins chapter.

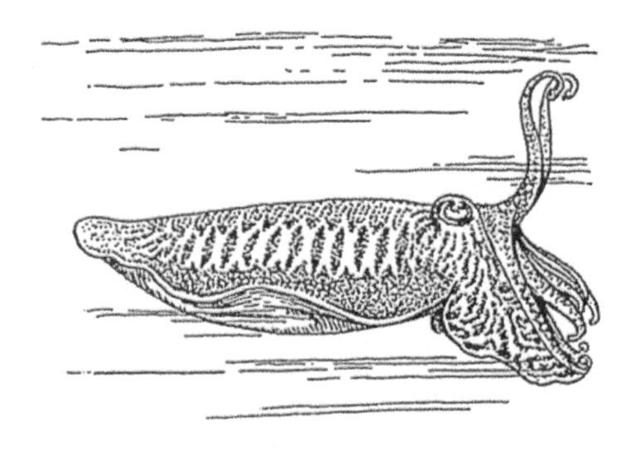

The type of prey depends on the habitat of the animal. Fish are the main diet of pelagic species (oceanic swimmers) and crustaceans are the favourite dish of benthic inhabitants. Squid can take a fish and finish it off by severing its hea. All the meat is stripped, leaving the tail and guts, which are dropped. Octopods search out crabs and other benthic creatures. Capture and death is swift, then the octopus returns to its lair (either a hole or a mound of rocks) to consume its meal in peace. The inedible parts are discharged, leaving a pile of body bits outside. This is how an octopus' lair is found by divers and spear fishermen so they know where to catch the octopus. It's the octopus' meal that leads to the octopus being a meal!

All the cephalopods have variations on the same reproductive ritual: the male performs a mating dance, often above the female. A vivid light display is also common and acts to warn off potential competition from other males. Then the two animals entwine at the head region and the male delivers sperm via a specialized portion of an arm. This has smaller suckers that hold the bundles of sperm to be delivered into the female.

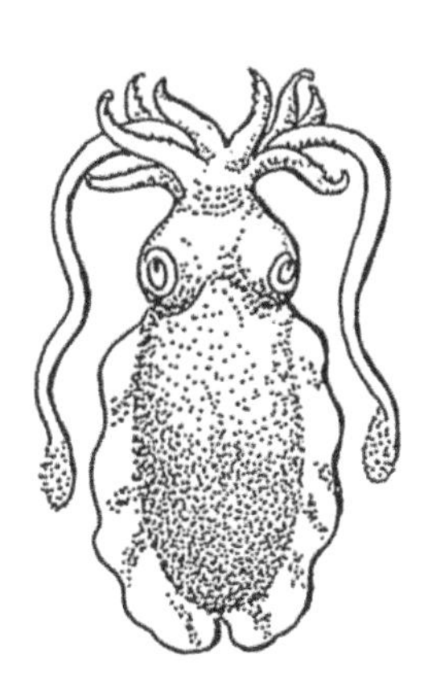

Once fertilized, the eggs are laid in bundles on the bottom of the sea, often in small crevices where the parents' arms can reach but animals looking for an easy meal cannot. Others species that perform mass matings lay a mass of capsules covering the bottom. The eggs are protected by a case that is not eaten by other beasts, since they either do not recognise this highly nutritional mass as a meal or chemicals deter them from taking a bite. Species that are pelagic lay planktonic eggs, often with specialized features to protect them. One such species secretes a calcareous bi-valved shell in which the eggs are brooded. Many species of squids and octopuses have short lives and die after the first spawning or in brooding species after the eggs hatch. The mating cycle is controlled via hormones which, when active, cause the animal to stop feeding, so that after the job is done, they die from starvation. One experiment involved a brooding octopus that had laid the eggs and was waiting for them to hatch. The animal had the hormone-producing gland removed and started feeding again; thus, the lifespan was increased.

Animals with Exoskeletons - The Crustaceans

These belong to the major grouping called the arthropods, which, when all its subgroups are combined, have a total of over 750,000 species, with many more certain to exist that remain unknown to science. This is over three times more than all the other species of animals known to science that inhabit the earth. Commonly they are known as insects, beetles, spiders, etc. However, the ones we are interested in include crabs, lobsters, shrimp and barnacles. The crustaceans have over 42,000 species and are mainly marine, but a few exist in freshwater and on land, most being permanent members of the plankton.

The two main features are the hard exoskeleton and segmented body, which go hand in hand as the exoskeleton is broken into segments to allow movement. Each piece of cuticle (exoskeleton) is joined to the next by an articulate membrane which stretches as the joint moves. This provides a barrier, protecting the inner soft, otherwise exposed body. This protection keeps out foreign bodies and parasites that are always present, waiting to invade a host.

Segmentation generally has three major regions: the head, thorax and telson, or the head, body and tail. Through the process of evolution, segments have fused together and been enlarged or reduced in size, often making the distinction between body segments difficult. In lobsters the telson is easily seen and is a major part of the animal, but if we move up the evolutionary tree to crabs, the telson is not evident. Looking under the animal, you will see that it has become reduced in size and curled round underneath the body. The female uses this as a protective area when carrying eggs that are deposited here.

As it is mainly inorganic, the cuticle that encases the whole animal (it's like wearing a medieval coat of armour) poses a serious problem. All animals grow—isn't that surprising—so they can only achieve this by shedding their skin. It's a case of 'off with the old and on with the new'. The old cuticle splits and the animal climbs out, instantly swelling to its new size. Now the animal is vulnerable to attack because it is soft, so it hides until the new cuticle hardens, which can take from a few hours to many days. The Antarctic krill probably shows the most rapid moult; if disturbed, it can shed its skin in seconds. There is no need for it to hide to allow the new skin to harden, as they exist in massive swarms often kilometres wide and long; protection here is in numbers. Each species has a limited number of moults within its lifetime; some moult every day; others, once a year. When the last moult has been reached, any damage that occurs is permanent. Up to 85% of its lifetime is spent fighting, so this last moult represents the beginning of the end for the individual. To illustrate how useful moulting is, if a shrimp loses a claw, it's replaced (often with the same size claw) on the next moult, but if it does not moult anymore, it's lost for good.

We now have an animal that's covered in cuticle and segmented. But what about the legs? These exhibit a great specialisation from species to species: some can only be seen by a microscope, but others are huge robust claws. The longest legs belong to the Japanese spider crab, which can grow up to six feet long. They are utilized for feeding, cutting up food, defence or attack, walking, swimming, filter feeding, stabilization, communication and copulation, in which the legs hold on and using specialized areas sperm is delivered to the female in small packets.

There is also a very important evolutionary process that has occurred with these beasts. The body shows a high degree of cephalization; that is, the formation of a distinct head region with a concentration of sense organs like eyes and a relatively complex centralized brain. This gives the animal an awareness of the direction of travel to nearby stimuli and enables rapid reaction to these stimuli. This can mean catching a meal or avoiding becoming one. The presence of eyes also allows communication between individuals and has allowed complex mating and territorial behaviour to exist. This has been shown as an effective method, not only to progress the species, but avoid unnecessary fighting, thus increasing the life span of an individual.

The Decapods- Crabs, Lobsters and Shrimps

The grouping Decapoda is only a small grouping within the crustaceans. Their distinction from other groupings are the legs. The decapods have all their legs attached to the thorax, and three pairs have been reduced and specialized into mouth parts, cutting up food and passing it for ingestion. The remaining 10 legs, five on each side of the animal, are easily seen. Their main function is locomotion, but they have important secondary functions, depending on the species and habitat. Ten legs are visible, hence the name decapoda.

With over 4,500 species, the crabs are the biggest group within the decapods. They are found in every environment within the sea and also represent the most recent animal to invade the land from the sea. Everyone knows the tale of how fish evolved into amphibians, crawled out of the sea and then became reptiles. The modern day equivalent is the crab, which can be found anywhere from the rainforest canopy to deserts, although most commonly, land crabs exist on damp forest floors. But they have one big problem that hasn't been rectified by evolution (a few more million years and they should be OK): they have to find water to reproduce, as all crabs have larval stages. At breeding time on Christmas Island, the whole island becomes a seething mass of crabs walking to the beach from the forest to breed, and when the young return the problem is even bigger as many more millions are present. Roads have to be closed because driving could result in human injury from sliding on the juicy parts of crushed crabs. The coconut crab also causes serious concern for the local human population. As its name suggests, it feeds on coconuts, breaking the shells with its claws to consume the soft inner nut. These crabs grow to quite a large size and are well known for their bad temper: they are extremely aggressive, which coupled with a formidable set of claws, makes them quite capable of severing an arm without a second thought! So be warned: leave them well alone.

The body of a crab is well known and characterized by the wide flat carapace, which covers the entire upper body, the legs protruding from under the side regions. The head area is located at the front of the beast, and the mouth is protected by the front legs, known as maxillipeds. These can be seen opening and closing when feeding in a tank; try this with a piece of mussel, but be patient because they are shy creatures.

The eyes are located on stalks that can be moved and turned in any direction. This allows the crab to observe any predators from any direction, including birds from above. The first pair of true legs are often topped with claws, whose shape is often an indication of the crab's feeding strategy. Fiddler crabs show great modification, with one claw the size of its body and the other reduced. This crab's large claw is used in fighting other males in territorial battles, and when breeding, it's waved around to attract a female. The other four legs are mainly used for locomotion. The majority of crabs walk or crawl over a surface, but swimming crabs have the last leg flattened to act as a paddle. A few years ago when in Thailand, I was taking a little dip in the sea, which supported a heavy sediment load that prevented me from seeing my body. I couldn't see the crab, but I did feel those legs crawling over my chest. It was only a swimming crab, but as I was young then, I certainly didn't wait around to find out which species.

Crabs make a meal for many predators including other crabs, octopi and fish, and their claws are only a minor inconvenience for these hungry creatures, so other, often novel strategies are used to deter any animal wishing to make a meal out of them. Some species place anemones on their claws and wave the stinging tentacles at a potential predator. The anemone obtains nutrition from broken-off body bits whilst the crab itself feeds. Hermit crabs carry these around on their adopted shell and when changing house, they also move their symbiotic squatters. The decorator crab, as its name suggests, places anything it can find on its body. It has flexible hooks that hold on to anything including bits of sponges, shells and stones. They remain inactive during the day and are very difficult to locate by sight alone.

Every feeding strategy that you can think of is utilized by these animals, from active hunting to filter feeding, and the shape of the claws always gives a clue to each species' preference. Most are active hunters and scavengers, but a preference for one prey group is always shown. Some species prey on mainly echinoderms, others mainly shell fish, and still others eat worms. This allows many species to coexist whilst not competing for the same food source. The prey taken can often be identified by claw shape. Species that eat shellfish have one claw heaver with blunt teeth that is used for crushing, whilst the other is slender and used for cutting flesh. Deposit feeders and algal scrapers often have flattened, spoon-shaped claws that are utilised for snipping off algae or spooning up sediment. Filter feeders have modified maxillipeds or antennae that stick up into the water current to collect particles, often whilst the crab is completely buried within the sand.

The reproductive behaviour and biology of decapods is quite complex but always sexual; never asexual. First of all, to reproduce you have to find a receptive partner. The emphasis here is 'receptive', because if a crab makes a mistake, he could quite easily lose a part of his body. Many species use body language and elaborate courtship dances. Fiddler crabs wave their enlarged claw to attract females, and different species have different waving patterns. When the female is attracted, the male bangs the claw on the substrate to produce a species-specific sound. Species of tropical ghost crabs build pyramids out of sand to attract females; the size of these pyramids is supposedly a reflection of the crabs' reproductive prowess. This usually occurs just before the female is ready to moult, though some species court just after a moult. After the ceremonies have been completed, the male grasps the female and climbs on top, this is to protect her, as copulation occurs just after moulting when she is soft and vulnerable. At this point I would like to suggest that if you see two crabs in this position, it's wise to leave them alone because the male is very aggressive at this time.

The sperm of decapods lacks a tail and thus cannot swim to find an egg, so it's packaged up into small parcels and deposited via the legs into cavities within the female. The female, after moulting, then hardens and the male crawls away, his job done. Meanwhile, within the female, eggs are passed individually past the sperm packet and fertilized. These are then passed to the telson, under the crab's body. Here they develop, and when hatched they join the members of the plankton.

Lobsters have a well-developed body plan even though they are below the crabs in evolutionary terms. Their telson is well developed, and the largest species attains a weight of around 20 kilos. They live in holes in reefs or under boulders and lead a solitary, nocturnal life; if you locate one, it's best to leave it alone unless you are skilled in handling them. Their downfall is the tail region which is highly muscular. This part of the lobster is consumed by humans to such an extent that many fisheries are seriously overfished.

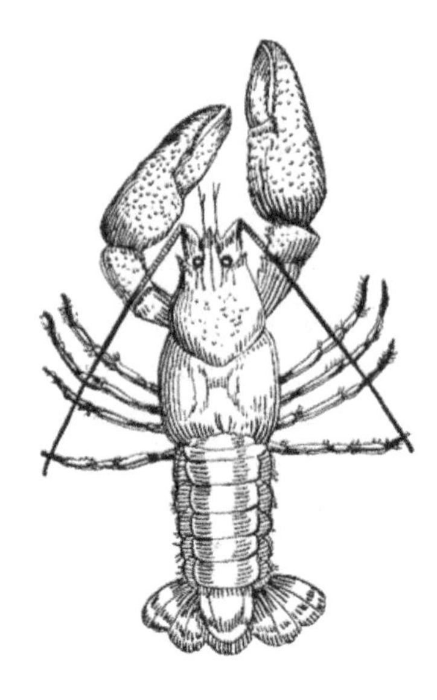

To meet this demand, many areas have tried to culture them, but there is one big problem: they are cannibalistic. If lobsters' young are not separated, you will end up with one big lobster. One study involved growing them separately to 6 months old, and then taking them out on to a reef and sliding them down a tube to a waiting diver, who then placed them in a hole. They are territorial animals, so after 6 years the survivors would still be in the area. The return catch was about 15% of the release figure, which is a good survival rate, but it was not good enough to continue. Lobsters are vulnerable to a wide range of predators when young, but as their size increases, their heavy body armour really takes effect and reduces the number of animals that have the ability to eat them. Their mating is similar to the crabs but, shall we say, they adopt a more human position, and the act of copulation does not last long.

Shrimp and prawns inhabit all areas of the seas, from rock pools and intertidal sand flats to the deep sea, some being totally oceanic and often performing feeding migrations of over

800 metres to the surface and down again in one day. It's best to think of these as small lobsters. The best way to tell if you have caught a shrimp or prawn is to employ the saying 'shrimps like sand, prawns like pools'. Take a powerful torch down to the rocky shore one night and go rock pooling; prawns' eyes light up in the dark, so it's a most rewarding experience.

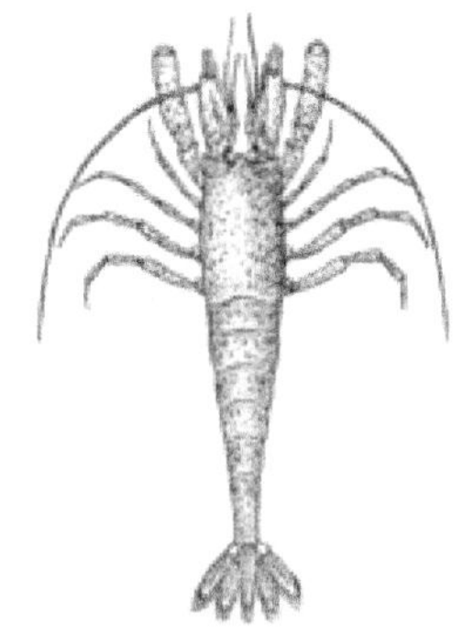

They feed on all sorts of food, mainly filter feeding and deposit feeding; however, a few specialised adaptations have occurred. You might have heard of the cleaner wrasse, a fish that inhabits coral reefs and cleans ectoparasites off other fish. Many species of shrimp do the same; these are highly coloured creatures that respond to a dance by fish at a local cleaning station. The fish remains totally motionless whilst the shrimp climbs all over its body, often poking the claws into delicate areas such as gills. If the fish is small, the weight of the shrimp drags the fish to the bottom, and it looks like the shrimp is eating a dead fish.

Though not a true shrimp, the mantis shrimp is a very powerful predator indeed The main beasts on its menu are crabs. As the name suggests, this shrimp resembles the praying mantis insect by having two large arms curled under the body, these are pushed forward and used to beat a crab. The force of this action has been measured and is likened to being hit by a bullet: the crab is beaten to death and the force of the blows shatters its body so that the shrimp can consume the juicy bits. Shrimp mate like crabs do, but hermaphroditic shrimp do exist.

Penaeid prawns are a very important shrimp in aquaculture; they make up the largest harvest of all species that are cultured, and often support local communities in under-developed countries. The world harvest is around ¾ of a million tons. That's a lot of prawns, as the average weight when harvested is 20 grammes. With advancements in science and in understanding the biology of different species, more and more species are now able to be cultured. Many species are being depleted at a vast rate for the Far East market, and these would soon be on the endangered list. Thanks to culturing, their future is now more promising. Also, whole areas of coral reefs have been destroyed in the past by cyanide and dynamite fishing for the aquarium trade. We can now culture over 35 species of fish for this trade and some of the cleaner shrimp too. Not only does aquaculture produce fish and invertebrates at a low price (salmon is now the price of cod); it also has the potential to save many species from the endangered list, and whole reefs from destruction.

The Barnacles

Why bother to look at a grouping of animals so common as the barnacle? Because they are amazing animals. The following paragraphs will reveal all.

One Mr. Charles Darwin was out to dinner and during polite conversation (this was after the publication of *The Origin of Species*) was informed that he knew nothing about barnacles. This remark must have hit home, because during the next eight years he devoted himself to studying barnacles. Think of this: he was ill then, and he spent the morning vomiting, the afternoon studying barnacles and the evening vomiting. He eventually produced two volumes dedicated to the humble barnacle that are still used as a source of reference today, and every reputable marine library has copies. After these were published, he referred to them as 'his blessed barnacles'. If Charles Darwin could spend 8 years studying them, I am sure that a few paragraphs are worthy of your attention, and you will never see barnacles in the same light again.

Barnacles belong to a group known as the cirripedia. The name is taken from the fact that their legs have become so modified into feeding structures that they are not legs anymore, but cirri. Both 'cirri' and 'peds' are legs; thus we have the grouping 'cirripedia' which are exclusively barnacles and contain their 900 or so species. These are the most highly-modified marine crustaceans and are effectively headless, with no abdomen or segmentation. The size range is a few millimetres to 75 centimetres.

They are found in most marine environments—and I really mean most environments—they are present in huge numbers in 'normal' habitats, often reaching a density of over 90,000 square metres of rock surface. The oldest species of barnacle alive to date was found at a hydrothermal vent site on the Mid-Atlantic Ridge, dating back 260 million years, making it one of the oldest living creatures today. However, many are parasites and can be found in such places as diverse as dolphins' teeth, corals actively feeding off the polyp and if you cut open a certain species of turtle (but please don't do this) you will find 200–300 barnacles attached to the throat. They cover many hard surfaces, and can be found on the backs of whales and crabs. They have caused millions of pounds to be wasted by their behaviour, covering pipeworks and ships' hulls.

There are two different forms of non-parasitic barnacles: stalked forms and non-stalked forms, which are more advanced. In non-stalked forms, evolution has acted to reduce and eventually eliminate the vulnerable stalk. The beast lies on its back and cements itself to a surface by a bio-adhesive, then builds a shell around the body. It still has a cuticle exoskeleton which is moulted allowing any damage to be repaired and growth to occur, but the soft new cuticle is not a problem as the shell provides cover to the animal when the new cuticle hardens. The legs are used as feeding appendages, but they also force water down into the shell, providing a current and allowing the animal to breathe. If you place a small barnacle-encrusted rock into a tank, you will see how these limbs beat to catch food. If you are lucky, you will see a mating. Also watch how the animals respond when you pass a shadow over them, for the shadow could be a fish waiting to nip off one of their juicy arms. The legs are coved in stiff hairs that give the appearance of a feather. These catch the food, but also force smaller animals, which could pass through the filter, down to the mouth.

So why don't the smaller animals just pass through the filter and escape? If you are a microscopic member of the plankton, water is a very different substance from what we humans perceive it to be. It's all to do with size and the weight of the animal; this was worked out by Osborne Reynolds, who was an esteemed physicist. We biologists are known to avoid mathematics at any cost and where possible, we leave that to other learned people. So Reynolds came up with a number for each sized particle. This is known as the Reynolds scale: an object's Reynolds number indicates how fast a particle will sink. To make a long story short, if you are a microscopic animal swimming in seawater it's like a human swimming in warm pitch. Filter feeders can catch food just by creating a current towards the mouth parts.

These beasts have one big problem: to reproduce sexually, they cannot move to find a mate. Most are hermaphrodites, but the male part still has to find another animal in order to deliver the sperm. They do this with an elongated penis, or multiples male reproductive organs known as pene. One species inhabiting northern waters has 8 pene, each 15 times the length of its body, the largest organ of its type relative to the body size. When the animal is feeling a bit horny, it throws the penis out of the shell and just prods around, searching for a receptive beast. This is the clever part: on this organ are sensory cells which pick up chemical signals from the receptive animal. These chemicals guide the penis to the part of the body that is waiting for the sperm's delivery. Variations do occur where animals sit on top of each other, a large female on the bottom and the smaller ones on top dwarf males waiting to attend the female. This is taken further in another species where the male attaches to a female. It stops feeding and metamorphoses into basically a sack producing sperm until all the energy reserves are used. It then dies and another takes its place. Oh, how good it must be to be the female of that species.

Eggs are brooded and larva called nauplii are released. These can remain in the plankton stage from a few days to a few months, and they develop through 6 stages. They then form a distinct non-feeding cypris larva. Now the countdown begins! This baby beast has to find the right place to settle and metamorphose into a feeding juvenile before its energy reserves are used up, or it will die. The animal doesn't settle anywhere, as this would likely lead to death, either by being eaten or by being crowed out by stronger animals. It recognises chemical signals from the adult population and thus is attracted to a safe place where it can build its home, grow and reproduce.

Animals with spiny skins - The Echinoderms

These animals are probably the most well-known marine invertebrates, and they have a fossil record dating back more than 500 million years. They are only found in the marine environment, in all seas, at all depths, inhabiting soft and hard substrates. Their members include starfish, brittle stars, sea urchins and sea cucumbers, among others, including over 6750 species. The best place to find them is either through diving or looking in rock pools or around the low tide mark, as they cannot tolerate much aerial exposure.

Be careful not to damage the animal when lifting; they can often be lodged into crevices, and tubefeet, spines and legs can be broken off if excessive force is used. Also be careful of the spines, as you may well end up the worse off, especially in handling a poisonous one. **When examining any marine animal, if you are unsure of their potential toxicity, *observe*. Don't handle**. Always place an animal into a shallow plastic tank, so water suspends the appendages; you will see a lot more than just looking at an animal held in the air. This is a basic rule for looking at ANY marine animal. Also, a clear-bottomed container allows you to see what's going on under the animal, which is often more rewarding. Replace the water often, as it soon heats up, causing adverse reactions from the beasts.

Characteristic Features

All echinoderms have spines on the skin. These spines may be long, short, thin, thick, microscopic, blunt or sharp. Whether you can see them or not, they are always there. Functions of the spines include locomotion, excavation, production of feeding currents and protection, as some spines are connected to toxin glands.

An internal skeleton is present, made up of calcite crystals. These may be connected together as in the sea urchins, producing a solid skeleton, or separately as in starfish, allowing flexibility in the legs, they exist as one crystal per cell. These skeletons give a robust appearance and provide protection from predators and wave action.

Specialized tube feet are present, performing a wide variety of roles and often being utilized in more than one way. Some uses for tube feet are locomotion, feeding, respiration, burrowing and sensing a wide range of stimuli. All tube feet are highly maneuverable, under hydrostatic control and able to be manipulated individually. Tube feet are located within the ambulacral groove running along legs or the body. This groove confers protection to the feet, as it is protected by spines that close over the tube feet when they are retracted. These features are easily seen when lifting starfish.

There are often many hundreds of tube feet per animal, each connected to a tube system known as the water vascular system. Water enters and leaves the animal through the madreporite, a porous, calcerous plate. This is best seen in starfish. If you look at the top of the creature's central portion, you will see a light-coloured area at the base of one arm where water enters the body. Here it passes through a tube system (like blood in our veins), running along each groove. At the base of each foot is a valve, connecting the foot to the system.

Through the open valve, water enters the foot and extends it. This movement is under the control of the nervous system. Water leaves the foot and the valve closes, allowing the foot to shrink and return into the groove for protection.

Over the body, pedicellariae are present, flexible rodlike projections with a formidable pair of calcite pincers on the end. Their function is mainly to keep the animal clean. To a larva, the hard skin represents a solid surface to settle and grow. If allowed, they would smother the animal, but they are easily removed. Specialized pedicellariae include large jaws for the capture of small fish, with a connection to a toxin gland. In most species to observe these, a powerful hand lens or low-powered microscope is necessary.

The Starfish

Most starfish have 5 arms and are around 10 to 20 centimeters across; however, the sun star can have over 40 arms, and other species grow to be the size of a dustbin lid or as small as a little finger nail. Most are active predators feeding on a wide range of food items: from scavenging dead bodies to digging out shellfish, they'll eat anything the can really settle on, as the mouth is situated on the underside. They are found on all substrate types, and have evolved specialized tube feet for each habitat. You will see suckered ends on those found on rocky shores that are used for attachment and feeding. Soft sediment species have spade-like ends for digging to feed or escape predators. On the tip of each arm are sensory tube feet that are longer than the normal ones and have no specialized ends to them. These tube feet wave around picking up chemicals to detect prey and other stimuli. They have the ability to regenerate; that is, if a predator takes fancy to a starfish arm, as he eats it, the starfish retreats and simply grows another. It is not uncommon to find many individuals with small regenerating arms evident; however, the process is slow and can take over one year to complete, depending on the initial damage.

A starfish undergoes sexual reproduction through external fertilization: the individuals mass together ready to shed their load. When the time is right, the first release from one animal occurs, causing a chain reaction throughout the whole assemblage of starfish, and the water above becomes laden with sperm and eggs. There is normally only one breeding season per year, with a female releasing over 2 million eggs. Once in the plankton, the eggs develop into larvae stages and it can take up to two months to reach the juvenile stage.

Starfish also exhibit asexual reproduction by splitting down the middle to produce two halves. The rest of the body then grows and a clone is produced. They are also able to reproduce in this way without their whole body. Some species can regenerate an entire starfish from just one arm. This kind of asexual reproduction is a very good strategy if the rest of it has been eaten.

Starfish have a very wide range of feeding behaviors, some ingesting anything they can subdue and others eating only one species of prey, often another starfish. Rocky shore species often take a fancy to bivalve shellfish (mussels). They grasp the shells with their tube feet and gradually open the shell to about one millimeter. The starfish's stomach is then everted through its mouth and into the shell, where it digests the flesh. Sandy bottom dwellers often detect their prey buried in the sand and dig for their food. When located, they ingest the whole shellfish and digest it, spitting out the empty shell when finished. Those that prey on other starfish grab an unfortunate, and with the use of their tube feet, they rip off an arm or two and ingest them. The damaged animal is then able to crawl off and start the process of regeneration until another arm is removed. It's a hard life being a starfish.

The crown-of-thorns starfish, which is often seen to be the scourge of coral reefs, readily consumes coral polyps. At their normal density of around 12 per 100 square meters of reef surface, they keep the reef in a healthy state by not allowing one species to dominate. However, in recent years, populations of this animal have increased, resulting in many areas of Pacific reef being killed off. We don't know if this is due to man's activities or a natural cycle, but these outbreaks are always met with high publicity due to the adverse commercial effect it has on tourism: no one likes a dead reef. This is often disguised in a concern for the reef's survival, but the fact, the damage is temporary. Soon new colonies will grow on the dead skeletons, and the reef will recover. One species of coral is always left untouched: within its structure lives a small shrimp. As the starfish settles on the coral, the shrimp eats the tube feet and the starfish soon moves on.

The Brittle Stars

This group of animals include basket stars, brittle stars and serpent stars. These beasts are easily identified by the distinct pentagonal central disc with 5 thin arms emanating from the center. Arm length varies with species ranging from short and stout arms to very long, branching forms resembling a bramble thicket. There are over 2,000 species of brittle star. They can be found in all habitats within the marine environment, and are sometimes extremely abundant, carpeting the sea floor and covering extensive areas at a density of over 2,000 square meter. Many species also live in a symbiotic relationship with other animals (an animal is termed symbiotic when, for most of its life, its natural habitat is another animal), including corals, algae and even the inside of a sponge. The sponge inhales water, from which the star picks up food intended for its host. At the same time, the sponge confers protection from predators to the brittle star.

Size again varies, and since arm damage can skew measurements, it is recorded by measuring the disc diameter. Normally species attain a size of between 1 - 3 centimeters, but they can reach up to 11 centimeters. Coloration is dependent upon habitat, with great variations within a single species. Rocky shore beasts are often dark-banded, either black or brown, and sandy bottom dwellers wear a tan colour. One species in Mauritius inhabits black lava flows that extend into the water: these have evolved a total black coloration to blend in with their surroundings. It's possible that this species has exhibited many colorations over the years, but the individuals with a lighter colour are eaten quickly because they are easy to see against their black background, so their genetic information is not passed on to the next generation. This results in only passing on genetic information that codes for a black colour, so only the black stars remain.

Spines over the arms also vary in length due to the brittle star's feeding strategies; these are also used in defense and locomotion and for lodging into crevices. There are two rows of tube feet along the bases of the arms that are used for adhesion to a surface and in feeding. No pedicelleriae are present, and the madreporite is located on the underside of the central disc.

Brittle stars are often subjected to high levels of predation, but they possess a novel escape mechanism. This can easily be demonstrated (if you are cruel enough). They don't feel pain, so all you have to do is grab an arm and pull gently. They automonise, or freely sever, an arm or portion of an arm. It's better to lose a part of your body than your life! Once its arm is severed, the animal suffers little disturbance from its loss, and can behave quite naturally, able to move normally (though its feeding abilities may be reduced). The lost piece is then regenerated, a common process that is easy to see, since the new piece is often a different colour and/or thinner than the original arm. In offshore beds, it is not uncommon for the population to show 100% of individuals with regeneration in progress. Not only does this allow the star to live and reproduce, but it supplies a constant source of organic material to be passed up the food web.

A brittle star's locomotion is not facilitated by the tube feet, as in starfish, but by the use of the arms. They can move in any direction and have no single leading arm, like a spinning disc that can shoot off anywhere. They only move if they have to, and most species prefer to remain sedentary. However, all are capable of rapid movement to escape predators or locate a food source. Soft bottom inhabitants are either rapid movers always on the lookout for food or they are burrowers. The burrow is excavated by the spines and lined with mucus to keep it from caving in. The animal lives here in its little house, with one of its arms protruding for feeding. This lasts until it is bulldozed down by a large beast moving through the sediment, resulting in either it being eaten or it building another house. If you have problems with your neighbors, just think about this poor little sod. Rocky shore inhabitants use their tube feet to adhere to the surface, but maneuver over the rocks with the use of the arms. Such is the ability for these animals to remain stuck on or in rocks by their spines and tube feet that, if you try to force them out, all you will achieve is the breaking off of an arm. When species form aggregations, locomotion is only utilized when an individual is crowed out. Then it clambers over the others until it finds a space and settles. Here it links arms with its neighbors for stability to avoid being swept away.

There are many feeding strategies within this group, which ranges from active carnivores to suspension feeders. Suspension feeders collect particles and plankton from the water column. Often you will see animals with arms stretched upward or sticking out from a rock feeding. Food capture is by one of two mechanisms: either by direct capture or within a mucus web. Direct capture is where a sticky mucus is secreted over the tube feet and spines. Particles (I use the word particle since anything can be caught, from a sand grain to a food item) are trapped by this secretion, then passed to the tube feet, which roll the food down the arm and cover it in mucus. Other food is collected, forming a small ball as it moves. The result is that when it reaches the mouth, a large bolus of food, mucus and inorganic particles is ingested. There are no table manners here. Basket stars spin an elaborate web of mucus between the upraised arms (like a spider's web), which captures particles. The web is then drawn in, packaged up, and ingested.

Deposit feeders are also common. These can be identified by an apparent lack of spines over the arms. They feed on any organic material that is present on the bottom, ranging from small items, which they simply move over and ingest, to large chunks of flesh, which are ripped apart by the arms. That's why there are no spines, as they would cause problems, getting clogged up with pieces of rotting flesh. If you find one of these, dropping a chopped up mussel into the opposite side of its tank causes an instant reaction. The star picks up chemical signals from the meat and will locate the food within seconds, rip it apart and ingest it. Its speed is quite an amazing feature. You will also see the central disc expand as food is crammed into the gut.

The success of this group, which has over 2,000 species in existence, is a direct reflection on the reproductive strategies it employs. Asexual reproduction is evident, with a beast splitting down the middle to produce two halves with three legs each (the sixth leg is grown before a split). The rest of the animal is then grown. Sexual reproduction occurs via direct shedding of eggs and sperm into the water, and planktonic development. Protective development is also shown by some species. This is where eggs are fertilized and placed within a body cavity known as the bursa to develop. Embryos develop on the wall of this cavity, without food passed to them from the parent, until fully-developed juveniles crawl out. This example shows the difference in reproductive strategies. The first produces millions of eggs with often only 2 surviving. Though successful, the adult population does not attain the high densities shown by the direct developers, who produce less than 100 eggs but enable all to survive to juvenile stages.

The Sea Urchins

These are the most characteristic of all the echinoderms, as in all species the spines are readily visible. Basically, if you think of a starfish rolled up into a ball, you have an urchin. Most species have a size range between 3 and 14 centimeters in diameter, but the largest grow to 36 centimeters. They come in a range of colors from black to a vivid multicolor. The internal ossicles are fused together to form a skeleton known as the test, a structure that, after an urchin's death, can often be found in tourist shops. Spines are all hollow, fitting to the test in a ball and socket joint to allow movement. They are covered with upward-pointing barbs and can be regenerated if lost. They can be short and stout or very long and extremely sharp. They are pointed at a predator if attacked, though they are often useless against specialized beaks possessed by trigger and puffer fish. Crustaceans also take their toll, breaking the spines and test with their claws. Pedicellariae are present on the body with many types connected to poison glands. These are used in defense, cleaning and breaking up large particles. An American species is highly valued in the Far East as an aphrodisiac—well, not the whole animal; just its reproductive tissues. Attempts by the aquaculture industry to farm urchins has met with little success due to their low growth rates, but with the development of new food this is increasing.

Two types of urchins exist: first are the regular urchins, which we are the most easily seen. Irregular urchins are oval in shape and live a life buried within soft sediments.

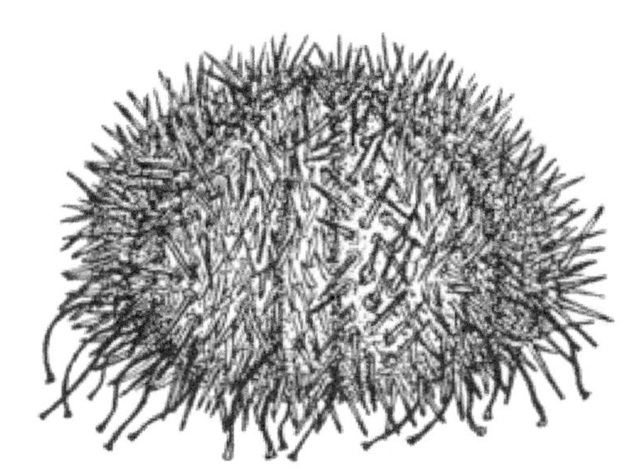

Regular urchins are round in shape and found on a hard substrate. They can be very abundant, especially in the tropics. The spines are longest around the center of the animal, decreasing in size at the upper and lower ends of the beast. Some species actively pick up shells, pieces of algae and stones and stick these to the spines to avoid predators. There are five rows of tube feet used for locomotion and adhesion to the surface. The animal's ability to move depends on the habitat and food source. With a rich food source, locomotion is reduced (why expend energy if you don't need to?) but locomotion increases when a food source is in short supply.

Animals exposed to heavy wave action are able to bore into solid rock by scraping at the surface with their jaws. Often whole burrows are created, with the animal excavating more as it grows. Eventually, it produces a large cavity within the rock with a thin opening that was formed when it was small. The result is that the animal is a prisoner within its self-excavated cell. However, it has its advantages: few predators are able to get at them and food is easily passed down the opening because of water movement. These species are very common worldwide. Coral borers often undermine the stability of a colony, resulting in it collapsing and falling to the deep sea floor.

If your food source is a microscopic layer of algae over an undulating rock face, and you are a round, ridged animal, how can you get at the algae and feed? With an Aristotle's lantern, which is the feeding apparatus of the regular urchin. This is a set of five jaws that floats on a set of muscles. As the urchin moves over the rock face, these jaws keep in contact with the rock and scrap off the algae. Others feed with specialized spines that have small pincers on the end. When a particle touches them, they snap shut—and hey, presto!—a meal. One species in North America feeds on kelp. The kelp forest is also the habitat of lobsters which take a liking to urchins. However, we humans prize the lobster, causing it to be overfished, which resulted in a proliferation of urchins, since nothing was eating them. The urchins consumed the kelp to such an extent that the habitat of the lobster was destroyed, so even after a ban on lobster fishing was implemented, they still have not recovered. However, help is on its way in the form of another urchin that feeds on the urchins that are eating the kelps, and so the food chain goes on and on and on.

The irregular urchins are commonly known as sand dollars or heart urchins. Their shape, instead of being round, has evolved into an oval, allowing easy movement through the sediment and rapid burrowing to avoid big beasts that might take a fancy to them. They are covered in small spines that are angled to the posterior end, used for burrowing (locomotion) and as a coat to keep the body surface clean. They have no Aristotle's lantern; they feed by either consuming sand and removing the organic covering and animals that live in between the sand grains, or by suspension feeding: protruding part of the body out of the sand and picking up particles from the water column.

Reproduction is the same for all species, that is, through planktonic development, though a few species show protected development. Burrowing species have long tubes which are passed to the surface, with eggs and sperm being delivered through them and released via chemical stimuli.

The Sea Cucumbers

These have no arms and can be thought of as a stretched out urchin with no spines. There are around 900 species of sea cucumbers, growing anywhere from only a few centimeters in length to longer than a full meter. They are extremely abundant in the tropics, but are found in all areas, especially in the deep sea. At 4,000 meters deep, they make up 50% of the total biomass, and at depths of 8,500 meters they constitute up to 90% of the total biomass.

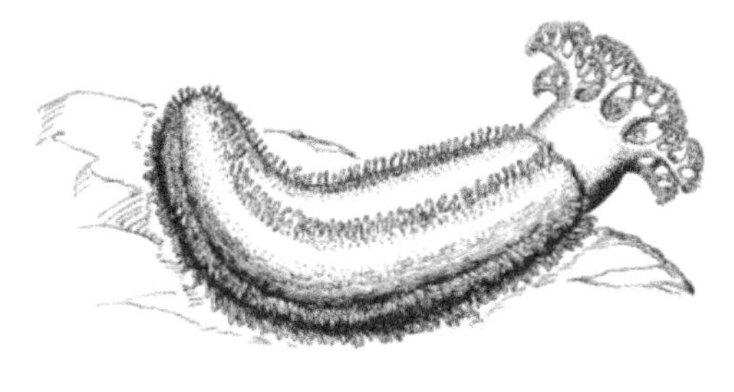

They have been likened to herds of buffalo rampaging over the abyssal plain. The internal ossicles are reduced to a microscopic form; however, some may protrude out of their leathery skin and act as hooks holding on to the sediment. The five ambulacral grooves are present: three on the ventral surface and two dorsally located. The tube feet here act for locomotion and righting the animal if it is dislodged. The only thing that is certain in biology, is that there is always an exception to the rule: in this case, one species has tube feet all over its body! The madreporite is internal and thus gains fluid for the working of the tube feet from the body fluids.

Locomotion is always with the aid of the tube feet, though muscular waves through the body can also aid movement. Most sea cucumbers either crawl slowly over the surface or burrow, but specializations have been developed, mainly in deep-sea species. Here tube feet can be vastly elongated to give the appearance of a cucumber elevated on stilts. Some have been fused together to act as fins and sails, allowing the beast to rise off the bottom and actively swim.

The mouth is surrounded by a number of enlarged tube feet that are often mistaken for elaborate tentacles. These can be fully retracted, along with the mouth, into the body. They pick up particles from the sea floor or catch suspended particles. Cucumbers can also act as sediment shovels: they open their mouth and move along, passing sediment through their guts. After the organics are removed out the remaining sediment comes behind them. These beasts are very important in re-working the sediment; it was estimated that, in a 1.7 square mile area off Bermuda one species turned over 500-1,000 tons of sediment per year. Basically, one adult consumed 120-140 lbs of sediment in a year. That includes some eating and recycling of the elements for other organisms to utilize.

If you have been unfortunate enough to have a cucumber eviscerate over you, it is surely an experience you will never forget. Evisceration is a very effective predator escape mechanism. When handled (divers, beware!) or attacked by an animal, the cucumber points its anal region at the attacker or handler. The latter half of the body ruptures open and all its internal organs located here, along with specialized tubules, cover the attacker. The head end then crawls away to regenerate a new anal end. Some species secrete special tubules without splitting, but if this fails, total evisceration occurs. With this mess of organs, some may also be sticky, and toxins may also be present, often resulting in the death of the predator whilst the intended prey escapes.

Asexual reproduction is not common; most reproduce via planktonic development. A very unusual strategy applies to some species in which the eggs are fertilized within the body; however, the mechanism that causes this remains a mystery to science. When young are developed, the anal region splits open and out they crawl. There are also over 30 species that are known to exhibit protective development: the fertilized eggs are trapped by the oral tube feet and transplanted onto the body wall, where they develop. - this may be one of the factors involved in that success.

Coral Reef Architecture

Coral reefs, what jewels in the sea, what oases of life! But are all reefs the same? What structures are found there, what processes are at work that force these structures to be built in the way they are, what differences do we find across the globe, why do they grow where they grow, should a coral reef really be called an algal reef? The list of questions is endless, we shall try to answer some of them. Most people look at a coral reef in wonder and constantly express their amazement at the array of beautiful life that's found there. That's just the problem, for they don't look at the reef itself, but at its inhabitants. So here, in answering the above questions, I hope to bring to you the wonder that is a coral reef.

All coral reefs occur within the tropics of Capricorn and Cancer, with a maximum development in water of around 24°C; they need well-aerated, clear water to live. This is why reefs are mainly found on the Eastern side of continents, due to cold, up-welling water on the West. They have a total global coverage of around 600,000 square kilometres. This is just 0.17% of the total sea bed, but 15% of the shallow coastal seas where most life is found. Throughout geological time, there have been five major extinctions of corals due to environmental change. The current reef systems have been in existence for around 5,000 years, as that was when the sea level stabilized after the melt from the last ice age. This great melt started 10,000 years ago, and by the time the climate had stabilized, the sea level had risen 125 metres to today's current level. Many species of coral did not keep up with this rise and thus perished.

There are two major reef zones in the world that exhibit extensive differences in coral inhabitants, and in the ecology of the area: the Indo/Pacific and the Atlantic/Caribbean reefs.. This phenomenon is due to the spread of the coral larvae from a single area known as the centre of speciation, where the most species are found in Indonesia. What this means is that in Indonesian waters the corals reproduced and many larvae travelled on the oceans currents until they died or settled. This is the first stepping stone in the species spread; these corals grew and reproduced, sending out larvae to the next stepping stone, and so on. However not all species produced hardy larvae, and so these remain in that area; only the most resilient species have travelled the globe. That is why we find the same or very closely related species in Australia as the ones in the North Atlantic, and why the North Atlantic has 57 fewer genera of corals than Australia. Scientists measure such differences in the amount of genera in an area. Everybody knows what a species is; a genera is the groupings of like species. There can be only one species in a genera or well over 100.

The table below shows the distribution of coral genre related to their location.

Location	Genera
Australia	71
South Atlantic Reefs	46-25
Borneo, Indonesia	67
(Most coral species exist in this location)	
Central Caribbean	25
Mauritius, Maldives	65
North Atlantic Reefs	14
Red Sea, Persian Gulf	46
Fiji, Tonga, Hawaii	13

Here we can see the species richness of the corals and how over the last 5,000 years, the larvae of these corals have travelled, colonising different parts of the world. It is evident that the Atlantic reefs are much younger that the Indo/Pacific reefs, which is apparent in the reef inhabitants of each area. When exploring, whether it be by diving or snorkelling over a Caribbean reef, the one thing that is evident are the large numbers of soft corals such as sea fans and also the large number of sponges that persist in very shallow water. These are relatively sparse over the Indo/Pacific reefs, only becoming apparent in the lower regions of the reef, below 50 metres. There is one very important structural difference that occurs between the two: the Caribbean reefs have no algal ridge, a point that will be explained later.

We have two differences on a global scale, but if we move down a stage, we find that there are three different reef types:

The first type is the fringing reef. These border the coastline in shallow waters and are separated from the landmass by a narrow stretch of water known as the lagoon. The lagoon can be as small as six inches wide, as in many parts of the Red Sea, or extending hundreds of metres like those found bordering Mauritius.

Second are the barrier reefs; the most famous of these being the Great Barrier Reef off the eastern coast of Australia, extending over 2,000 kilometres. The second largest is found off the coast of Belize in South America. These are often situated kilometres away from the shore, their base being the edge of the continental shelf and dropping off down to the abyssal plain. Expeditions which have travelled down to this base have found the area littered with coral structures often bigger than a house and weighing hundreds of tons, that were dislodged and fell to their deaths.

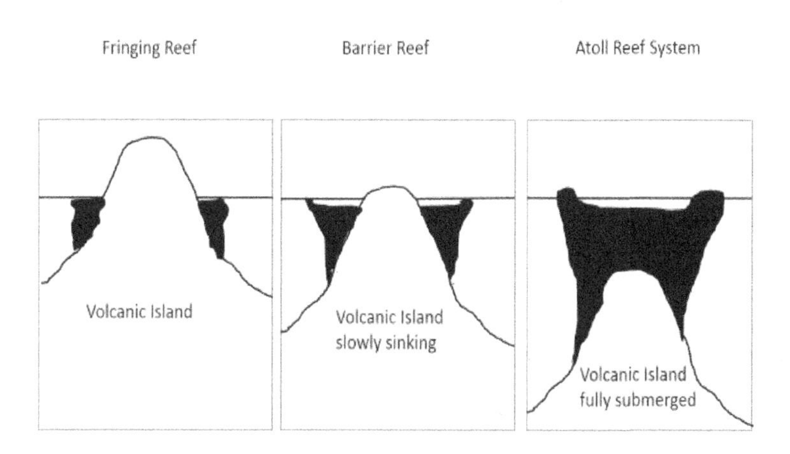

Finally come the atolls. These reefs are mainly found in the Indo/Pacific and owe their origins to hot spots under the Earth's crust. This is an area where the crust is thin and heat from the magma causes the crust to move upwards. Over hundreds of thousands of years, a new volcanic island is formed, like Hawaii. With the earth's tectonic plate movement, the island moves away from the hot spot, and at the same time a fringing reef is born around its edges. However, as the island moves away, the Earth's crust cools, becomes denser and sinks in deeper water, so the reef has to grow upward to remain in the surface waters. This upward growth equals the rate of sinking; after a long time the island eventually loses its fight and sinks beneath the waves, leaving a ring of coral reef around a central lagoon, and an atoll is born. All atolls were once fringing reefs, but not all fringing reefs will become atolls. This was first put forward to the scientific community by Charles Darwin, who postulated it on his famous voyage on HMS Beagle. This remained in question until the 1950s, when the first cores were taken from such an atoll. The coral limestone went down over 1,500 metres until it hit volcanic rock, when Darwin must have had a heavenly smirk on his face.

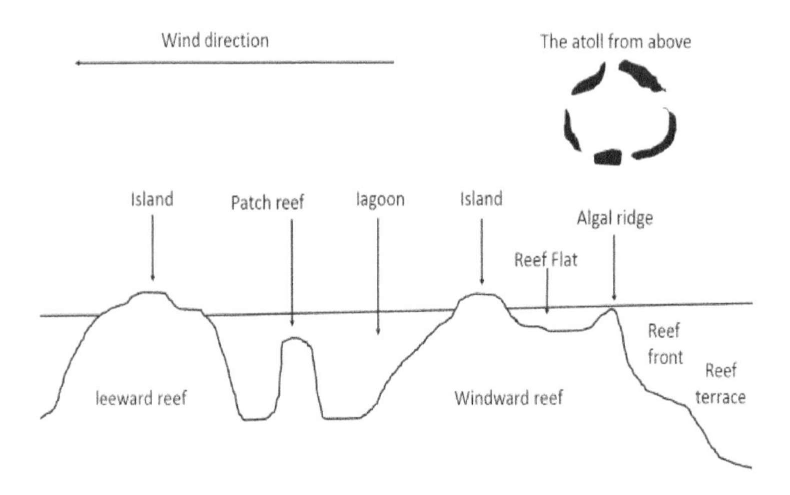

The formation of a reef has been studied by shipwrecks which have occurred miles away from any reefs, but the coral larva found them (a stepping stone) and settled, thus allowing scientists to study how a new reef is born, and how it develops. The succession on a reef has been shown to demonstrate a strict pathway.

The first stage that occurs on any hard substrate is that the surface is covered by a bacterial film. This is very important for any larva that has the potential to settle. The animal will not do so unless it can detect such a bacterial film. This is true for any larvae from any species of any animal in the marine habitat. Once such a film has built up, the first inhabitants are microalgae and large seaweeds. Then the motile and sessile invertebrates move in, including snails, barnacles and clams. The sessile animals are the key to a new reef, for the shells are where the coral larvae settle and start to grow. This stage takes about one year until we move into what is known as the primary phase.

Primary corals move in and start to colonise the area. These corals have very resilient larvae who will settle on virtually anything. If they get it wrong they die, and if they get it right they live; what a choice: it's Russian roulette for corals. These primary corals show a number of traits that allow them to be successful; they are the fast growers, and as such are the dominant species on a reef. This rapid growth causes a problem, because after a period of time the colony becomes unstable and they snap off, falling to the deep. They may die there, or they can grow if they fall in the shallower waters, cementing to the substrate. If a breakage does occur, for that section it's bad, but for the reef it's good as the newly exposed surface allows new coral larva to settle and grow. This primary phase lasts about seven years, as it takes that long for the corals to achieve the size that is required before the secondary and final phase starts.

In the secondary stage, we have many relatively large corals growing and they are now in danger of being broken either because of unstable growth, or from bio-erosion, in which countless animals of all sizes bore into the calcium framework of the coral, weakening it. So the secondary coral species move in. These are slow-growing forms; however, they grow in between the primary species and cement the framework together. Corals are particularly aggressive creatures, and the major resource that limits a coral's growth is space, so when two colonies touch the opposing sides do battle with their stinging cells, resulting in a dead area between two colonies. Nature provides many wonders, and here it has armed the slow growers with a very nasty temperament and powerful toxins, stopping the fast growers from crowding out the secondary species, for they are no match for them, and coral diversity is maintained.

At the same time, other animals settle in between the corals, such as giant clams and other molluscs. These are soon cemented into the structure, further strengthening the reef. Now we have what is known as a new reef, for it is still young and it must start growing as a whole. Some state that a reef can be considered a whole organism, growing and providing habitats for others.

Reef growth is dynamic: growth occurs, and at the same time erosion takes place, so growth has to proceed at a faster rate than erosion. The whole area is taking two steps forward and one step back, because when growth occurs new holes are bored out. Here the corals find an ally in an unlikely form: a bacteria. As a hole appears, the bacteria colonise it and wait, in the water column, many microscopic grains of calcium are floating around, and soon the hole will receive one; this is when the bacteria get to work. As the grain settles, bacteria cover its surface and produce a cement, thus filling in the hole and repairing it, only for it to be bored out again and filled back up at a later date.

Without these bacteria and the role they perform—a process known as submarine lithification—coral reefs would not exist, for this process results in the outward accretion of the reef slope and stabilizes the drop-off reef wall.

Many species of corals are able to assume different-shaped colonies: in one part of the reef, a single species may show a branching shape but further down the slope it might assume a tabular shape, like plates stacked on top of each other. This results in a zonation of not only species of corals but the shapes they grow in as we travel down a reef front. Many different factors cause this to happen, but we will look at the two main stresses exerted that have a profound influence on the overall construction of the reef front.

The first is hydrodynamic stress. As the ocean moves towards the reef, it comes into contact with a large barrier, the reef itself. Pressure builds up in the water column, eventually crashing as powerful waves against the reef front. Contained within these breaking waves is an immense concentration of energy that has to be dissipated. It can go two ways: over the top or down the front, which is the fate of most of this energy. This very powerful water movement cuts huge surge channels through the coral framework. The channels travel down to the bottom of the coral reef, either to the drop-off wall or sand plains. These surge channels effectively funnel all of the surface water down to the deep. They are often 3-4 metres wide and tens of metres deep. Along the sides we find brain coral colonies, as these have the most structural strength.

However, near the bottom of the channel, no life exists (that we know of), as the water velocity is great: combined with many sand particles, we have a naturally occurring sand blaster working at all the exposed surfaces.

The second stress on the reef is light penetration: the influence this has on coral shape is profound due to the fact that corals receive most of their nutrients from algae growing in the cells of the gut wall. The fact is, algae need light to live and function, so with diminishing light penetration the deeper we go, the algae are able to give less and less food to their hosts.

However, at the very surface, we have a high concentration of ultraviolet light. This form of radiation can kill the algal cells, so again we find that the reef has less coral here with different shapes. At the very surface at the back of the reef front in the shallow lagoon, we have mainly branching forms, because this intercepts less light but allows polyps to function normally.

Also, as this area is very calm, the branching forms are in no danger of being broken by hydrodynamic stress. There is little coral growth from the surface down to 5 metres on the front, as UV light penetration is dangerous, so the most growth occurs between 5 and 30 metres down. Below this, the lack of light becomes a limiting factor and as such, many corals take on a tabular shape, allowing the surface area to catch the maximum amount of light on the reef terrace. In the Caribbean coral, growth occurs down to 100 metres deep, where coral growth is slow and sparse and corals mainly rely on a carnivorous diet; in the Indo/Pacific, growth penetrates down to 60 metres, not due to light but to the crown of thorns starfish which lurks mainly at this depth. This animal is absent from the Atlantic.

The typical reef takes on this form:

The lagoon is an area mainly consisting of calcareous sand. Sea grass beds can be extensive, and in the deeper areas patch reefs exist. There are mainly branching forms of coral, but also brain corals and tabular corals in the deepest parts.

The reef back is the area directly behind the algal ridge: a sediment-laden area from the reef front, where many species of calcareous algae are commonly mistaken for corals. Giant clams and anemones are also common. The fire coral inhabits this area. This is not a coral, but a hydroid with powerful stings.

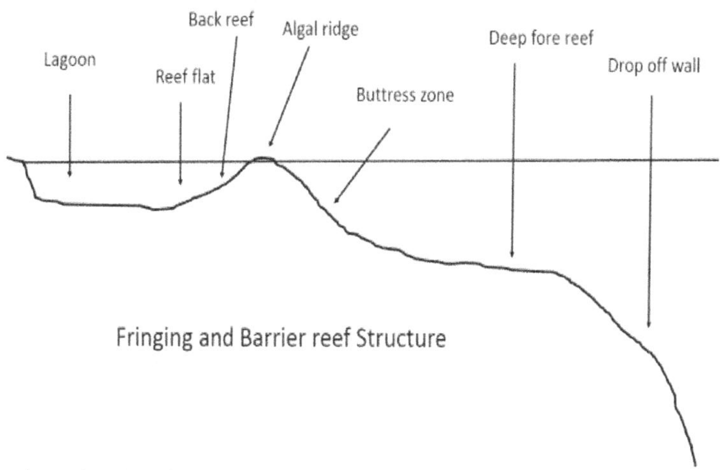

Fringing and Barrier reef Structure

The algal ridge is an area of exposed reef. There are over 7 different species of encrusting calcareous alga. This region looks like a cement wall and takes the full force of the breaking waves; cut extensively with surge channels, it extends down to 5-6 metres, and a few broken colonies of coral exist here.

The reef front is the area of highest coral diversity and growth, extending from 10 – 30 metres deep. The highest diversity of reef inhabitants are also found here. The structure is cut with surge channels, giving a spur and groove formation. This is a very high-energy environment.

The reef terrace extends 30 to 100 metres deep. In the upper regions, the main coral growth is tabular with less and less coral growth as depth increases. The main hard substrate is calcareous red alga. The increased space allows sessile animals such as soft corals, sponges, shellfish and urchins to colonise extensively. The region terminates steep drop off or sand plain. In one of the questions in the first paragraph, I asked if we should call coral reef algal reefs. If the naming of this area is taken from the most dominant life form, then the latter is true, for indeed there is more algae here than coral. In the 1950s a series of cores were taken down to 1,500 metres deep, and when analysed, the calcium-producing organisms were ranked. In fourth place came corals, in third came a microscopic animal, and in second and first places came calcareous algae. The question is, where is it on the reef? It's very hard to recognise or it just doesn't look like a seaweed, but the fact remains that 90% of the sand in the lagoon is produced by just one genus of algae. When we think that lagoons represent 95% of the reef system, we really think that they should be called algal reefs. But it's just not as romantic, is it?

However, when we postulate about such a statement we must move away from the romantic side of life and think about the biological implications to see how important this is. The clear blue waters in the tropics are nutrient deficient; therefore, they can support little algal growth and can only hold a small plankton population. That's why they are so clear. The basic fact is, without algal growth, there is no or little life, as these plants are the basis of 99.9% of all ecosystems. Without algae, there is no reef; when we have no reefs, we have no life.

In the lagoon area and over the reef front, huge amounts of filamentous algae grows, but as soon as it's grown, it is eaten, leaving only a small amount showing. In the lagoon, damsel fish actively farm their food: the fish weeds out any old, less productive growth and the different species that are not eaten.

This ensures a constant supply of new shoots to nibble at, and at the same time prevents the algae from smothering the corals. They will defend their farm and attack anything or anyone who invades.

Another area where algal growth is of paramount importance, when considering the reef as a whole, is the algal ridge and reef terrace. Here we have encrusting calcareous alga which builds a massive wall, taking all the wave action and protecting the lagoon, and further down, often below 30 metres, the algae takes over from the coral as the main provider of substrate, allowing other animals to colonise the area. This type of algae is really funny stuff indeed; less than 10% of each plant is organic, and the rest is calcium carbonate, which produces the cement-like wall.

We have seen how between little coral polyps and algal cells, huge calcium carbonate structures are built and maintained. So what do we know about the animals that inhabit such areas? The answer is very little. We know that around 60,000 species inhabit these areas, but calculations have shown that we should see about 423,000 species, so where are they? These animals are there but we haven't discovered them yet. This is mainly due to two factors: the first is that SCUBA has only been used since the early 1960s, and the second is that money is not available for such research. With immense areas still to be explored and studied, we are more likely to find a cure for cancer on a reef than in a rainforest.

What is the current state of our coral reefs today? A survey recorded that out of the 109 countries in the world that have significant reef coverage, 93 showed serious damage. If reefs failed, the consequences would be devastating both commercially for the local inhabitants who often rely on the reef, and ecologically, as the food chains would shift and many species would become extinct. So what damage does occur?

Enriching the sea with nutrients kills corals very rapidly. When we build hotels, the sand is compressed and all the organics move to the sea; fertilizers run off from the land due to agriculture; sewage input—all give nutrients. To the filamentous algae, it's a field day and growth is rapid, smothering the corals. Within one year, the whole community changes from a coral-based one to an algae-based community. The number of species found is vastly reduced as the alga covers every surface.

Deforestation in the tropical rainforest not only wipes out all the land animals, but with the rain comes soil erosion. This has two effects. The first is it reduces the amount of light penetrating the water column and stopping the corals from growing. Second, when the sediment settles, the coral polyps become clogged and die.

Humans also remove large chunks of reef by dynamite fishing, mining, coral and shell collecting, anchors and chains and the list just goes on and on. An average reef will grow one to four centimetres per year, but we can remove hundreds of years' worth of growth in just seconds.

If reefs fail, many areas would not exist anymore because of flooding, thousands of people would die in poverty from starvation, unemployment in areas where jobs that relied on reefs became non-existent would rise a hundredfold, and many divers would not get their holidays. When exploring these reefs, think how we can save them!

Marine Invertebrate Toxins

There are thousands of species of animals within the sea that produce venom and toxins. The vast majority are relatively harmless to man; most only cause minor rashes. The most famous animals are fish and, within the invertebrates, the box jellyfish and blue ringed octopus; however, the fact remains that, when compared, their active agents used are of a much lower potency than the few we will be looking at in this chapter. So here we will concentrate on some of the most potent toxins known and how they cause many poisonings and deaths each year. Also because of the nature of the toxins, these are indeed unusual and relatively unknown outside the areas they affect. The first question that needs to be asked may sound simple, but when thought about, it has a far-reaching meaning: 'What is a toxin?' In the broadest sense, it is a molecule that occurs naturally or is manmade and has an adverse physiological effect on a living organism. This definition is very loose, due to the fact that water can have toxic effects if drunk in extreme quantities. Native northern Canadians know not to eat the liver of a polar bear because if they do, they will die from vitamin A poisoning due to the very high levels of vitamin A within the liver. So we have to narrow down the definition. For our purposes, we will devise a scale starting with cyanide. Any naturally-occurring invertebrate toxin which can kill a man using a smaller dose, shall be considered as a toxin for this small chapter. On our scale, 10,000 units of cyanide will kill a man. In the table below are the toxins that we will talk about, their origins, their delivery route to humans and how much is needed to kill an average adult male.

Here we have two separations: venoms and toxins. Venoms are proteins which in molecular sizes are enormous. Toxins are very small molecules compared to them; it's like a 50-story skyscraper and a one-room wooden shack standing next to each other. Many people would claim to know all about these venoms, like sting ray, stonefish, weaver and the other 200 or so species of venomous fish. These often sting people who are unfortunate enough to tread on them. When they do, help is at hand in the form of antitoxins if you can get to a hospital in time. Most venoms will not be included in our table because cyanide is more deadly, but as a general rule anything that stings is a venom. All venoms are poisonous, but not all poisons are venoms. The most deadly toxins produced by invertebrates are not venoms and have no antidote due to the fact that they are a very small molecule. This prevents us from being able to develop a chemical that can bind to it and render it harmless. Oh dear!

Toxin	Delivery	Units to Kill
Cyanide	Ingestion	10,000
Conotoxin	Sting	1,000
Brevetoxin	Eating shellfish	166
Okadic acid	Eating shellfish	64
Enhydrotoxin	Sting	13*
Saxitoxin	Eating shellfish	3
Ciguatoxin	Eating reef fish	0.15
Maitotoxin	Eating reef fish	0.056
Palytoxin	Arrow tip (coral)	0.052

* *Vertebrate example used to compare toxicity with amount delivered (see paragraph below)*

From the above table we can see that cyanide is a relatively harmless molecule compared to what some marine invertebrates can produce; however, it is often not the potential toxicity of a substance, but the amount delivered in one go that is often the deciding factor in whether an animal will live or die. For instance, Conotoxin is far less potent than sea snake toxins, but if bitten by a snake your chances for living are 50/50 with no treatment, whereas if stung by a cone shell, without treatment, you will die within 4 hours. This is due to the amount of toxin delivered. Ciguatoxin poisons between 10,000 and 50,000 people each year, yet fatalities are not common even though it's over 10,000 times more deadly than cyanide. This is due to the amount ingested, but it's still a nasty little molecule and can cause serious illness for up to seven years. We'll meet this little chap later.

So how are toxins delivered to us humans? And how do they affect us, causing either serious illness or death? They are delivered by various means. The most common way is through the food chain, in which the organism producing the toxin is often a bacteria living within an algal cell. The alga is eaten and the toxin does not affect the consuming animal. This animal then stores it and accumulates the toxin to deadly levels, then we eat it! The other way is for a person to be stung. When this happens the, press get a hold of it: 'The Box Jellyfish Strikes Again' and other headlines appear, but the truth is that death occurs at a much lower rate through stinging than through eating infected animals. An example, if you eat a mussel that is toxic, you will be paralysed within a short time. This is from ONE mussel. It takes two metres of tentacles from the box jellyfish, full of thousands of stinging cells to kill, so this animal is much less dangerous.

How do they cause death? Most act on our nervous system, in one of two ways. One way is blocking the transmission of nervous reactions, causing paralysis of either the heart or respiratory system. In either case, you die of a heart attack or through lack of oxygen as you cannot breathe. The second way is to excite the nervous system, giving your body uncontrollable movements. The most common cause of death is by heart attack, as the heart goes into overdrive.

Some toxins are not nervous agents, instead acting by directly killing cells in the body. These often are transported through the body, killing cells very rapidly and thus killing us in a short amount of time. In all cases, incredible amounts of pain accompany you to your grave.

Limu-Make-o-Hana (the Deadly Seaweed of Hana)

On the Hawaiian Island of Maui lived a tribe of natives who throughout native history were undefeated in battle. During troubled times before a battle, the local witch doctor would take all the arrows and spears down to the seas edge at a holy place called Hana. He would perform a ritual, blessing the tips of the weapons and wipe them with a seaweed. In battle, if individuals of the opposing side were scratched by such a tip, death was agonising and rapid.

Marine biologists stationed on the island were intrigued by such stories as often, scientific facts are discovered by such folk law. In the early 1960s, when one scientist tried to get the tribe to reveal the location of the limu-make-o-Hana, he was warned off under a threat of a curse. However, being a scientist, such a curse did not bother him and so he kept on getting the witch doctor's assistant drunk until one day he was told of the location of such magical powers. The next day, he left the research station and found the seaweed, but he returned to find the marine station burned to the ground, and the witch doctor's assistant dead!

What he had discovered, however, is one of the most amazing things in marine life: the seaweed was in fact a new species of soft coral, which secreted the most powerful marine toxin known to man: palytoxin. The genera of this coral is well known and goes by the name of Palythoa, its species known all over the tropics and all toxic. This new species was something different, because the other species are relatively harmless compared to this nasty beast. The species was given the scientific name *Palythoa toxica* for obvious reasons.

Apart from producing Palytoxin, this soft coral is unique in two other ways. It was first discovered in a small rock pool about 4 feet across, above the high water mark, which means that this pool was never connected directly to the sea, and water entered the pool by either sea spray or rain. The salt content of normal sea water is 35 parts per thousand, but the rock pool had a salinity of only 25 parts per thousand, deadly levels for most corals. Therefore, this animal not only produces the most potent toxin, but it lives and thrives a rock pool deadly to most corals. Now for the amazing bit: this rock pool is the only place on Earth where this species of soft coral is found. Think about this for a minute: this is the only place it lives, so through asexual reproduction, hundreds of years of coral generations have remained here. Through this isolation, it has evolved host algae and bacteria within its cells to produce a unique, potent molecule. It's the marine version of Darwin's finches.

Many attempts to grow the coral in different pools around the island have been tried, but all have failed. To this day, that rock pool is probably the most biologically deadly place on Earth to mammals, as invertebrates are not harmed by the toxin.

Work on this animal has been slow due to the fact that the scientists have an agreement with the local tribe to remove only small amounts for study every year, allowing a good relationship with the locals because the area is still considered holy. In doing this, they also avoid endangering the coral colony by removing too much at one time. However, a few facts are known: the brown coral looks like a seaweed until disturbed, when it turns a purple colour and its tentacles retract. At the same time, a yellow substance is extruded from the animal and fills the rock pool with toxin. All the witch doctors had to do was disturb the coral and dip the arrow tips into the water, then allow them to dry, and hey—presto! One of the deadliest chemical weapons is produced. The toxin in fact is not produced by the coral itself, but by a bacteria which lives in a symbiotic relationship with the coral cells. This bacteria has been isolated, but only 35 microgrammes of pure Palytoxin has ever been produced in this way, but we now know how it's produced and that the bacteria lose the ability to produce the toxin when isolated. We also now know the toxin's chemical structure and how it acts.

Palytoxin has a cytolytic effect (it kills cells) so it moves through the body killing all the cells until it comes into contact with a major organ, for example, the heart. The heart cells are knocked off and the victim is killed. Pharmaceutical companies are interested in all types of naturally-occurring toxins but cytolytic toxins are important in the fight against cancer;

however, the trick is to produce a toxin that only kills cancer cells. Press coverage of finding cures from rainforests is extensive, but coral reefs are a more likely place to find such cures and most work is centred in this area.

Shellfish Poisoning

This is a type of food poisoning in which shellfish that have been exposed to an algal bloom are consumed by humans, resulting in various ailments depending on the species of algae blooming. The range of symptoms include an upset stomach, diarrhoea, vomiting, numbness in fingers, permanent loss of short-term memory, permanent paralysis and death. Not really a nice mixture.

But what is an algal bloom? When certain variables, including the right temperature, the right nutrient concentration and calm seas all align for a few weeks, this produces the optimum conditions for algal growth, resulting in an amazing division rate of cells. Within a week the concentration of algae can explode from a few hundred per litre to well over 5 million per litre, and it's very destructive for both animal and human health. Some species are red in colour, and blooms of these turn the sea red. Hence, the term 'red tide.' The sheer density of cells and the mucus they produce clogs fishes' and larvae's mouth parts, as well as respiratory structures such as gills, causing mass mortality. Often the first indication of a bloom is hundreds of tons of floating dead fish. When the bloom dies all the dead cells sink to the bottom. This can smother all benthic life (life on or near the sea floor) and cause anoxic conditions, resulting in the death of all benthic animals.

Here man has a problem, for the incidence of blooms is increasing both in frequency and duration on a worldwide basis. We don't know if coastal sewage disposal is to blame, enriching the nutrients in the sea, or if the blooms have occurred at this rate through ages of ecological time, with our technological advances in monitoring equipment simply allowing us to see more. Only time will tell. The agricultural pesticide runoff from the land has a double whammy effect on blooms, because not only does it increase the nutrient content, allowing such rapid growth, but it also kills off the small algal grazers, so the bloom grows unchecked.

So what happens when a bloom appears on a commercial shellfish bed? The effects have a far-reaching commercial damage. Apart from all shellfish being banned from sale, all other marine products suffered a drop in sales and price when consumer confidence dropped because of bad reporting. This can affect jobs in a wide range of support industries, such as marketing, transport, etc. and all because it was not stated that only shellfish are affected and all other marine produce, like fish, remain toxin free.

So what happens to the infected shellfish? This depends on the species of shellfish, as they store the toxins in different body parts, but with time all can be cleaned and sold safely. Mussels are the quickest to release their stored toxins, followed by scallops, with clams remaining toxic for the longest period. There are many ways to remove the toxins, involving various chemicals and expenses, but the most effective way is simply to transfer the stock to clean seawater and let the animals do it naturally.

Before being reintroduced into the market, extensive tests are performed to see the level of toxins present.

There are a number of different types of shellfish poisoning, each with different degrees of symptoms, but unfortunately for the infected, there is no antidote.

Paralytic Shellfish Poisoning (PSP)

A few decades ago in America, an outbreak of paralytic shellfish poisoning occurred, resulting in many deaths, and scientists were called in to find the cause. For sixteen years, PhD students went down to the shore every day and collected a few mussels to mash up and look for a toxin, with no positive results. A red tide occurred, and the next day, the mussels were full of a very potent toxin, but the alga did not produce it. It turned out to be bacteria living within the cells. Mussels consumed the rich food source—they can easily eat over 50 million cells per hour—thus storing the toxin and concentrating it to deadly amounts. The scientists were very pleased with themselves, and justly so, until a local Native American explained that they had known for centuries not to eat the shellfish when the tide turns red!

This is the most potent kind of food poisoning known to man, and is caused by blooms of at least seven genera of algae whose characteristics include massive fish mortality and luminescence at night. The principal toxin involved is saxitoxin, named after the Alaskan butter clam from which it was first isolated in 1957, but now at least 18 different types of saxitoxin have been isolated. The toxin acts on the nervous system. Its symptoms include dizziness, permanent paralysis, death from cardiac failure, and respiratory failure due to paralysis of the diaphragm. This type of poisoning occurs on a worldwide basis, and is caused by us. The following tale reveals all.

An outbreak of PSP occurred in Tasmania, an area that was previously free from such events. The species of algae that caused the poisonings was unrecorded in the Southern Hemisphere, so the detectives were called in. When a large tanker is travelling empty, it takes on ballast water to stabilize it in the ocean. The algae species was transported from the north to the south in this way and when the ship reached port, the ballast water was discharged and a new species was introduced. The actual ship that carried the algae was located months after the incident; a great piece of detective work by anyone's standard. From this example, it is seen how easy it is to spread deadly toxin-producing organisms across the globe, so we have to be careful.

Neurotoxic Shellfish Poisoning (NSP)

This is caused by a particularly nasty algal bloom, recognisable by a discolouration of the water, massive fish mortality, and a respiratory irritant in the air. The principal toxin is known as brevetoxin named after *Ptychodiscus breven*, the algae in which it is found. Symptoms are the same as PSP, but I am pleased to say, it is much milder. No fatalities have occurred and symptoms desist after 36 hours. There have been outbreaks in the Gulf of Mexico, the Caribbean and Spain.

Diarrhoeic Shellfish Poisoning (DSP)

This type of poisoning was only discovered in 1983 as the symptoms include an upset stomach and a case of diarrhoea. As such, DSP was often mistaken for gastroenteritis. However, more alarming is that a bloom is not required to produce infected shellfish. Algal concentrations as low as 100 per litre are all that's required to cause DSP—and this is a very common density. The toxin is called okadic acid. Thankfully, symptoms are short-lived and mild. Occurrence is thought to be worldwide, but mainly in the warmer parts of Europe, the US, Japan and Mexico.

Amnesic Shellfish Poisoning (ASP)

Only one outbreak of this condition has occurred—on Prince Edward Island in 1987—but it left devastating effects. After a massive bloom, 150 people consumed blue mussels containing a high level of domoic acid, which killed 4 and left the rest with a permanent loss of their short-term memory.

Most cases of shellfish poisoning are mild and not associated with algal blooms, and have symptoms similar to DSP. These countless cases are caused by a high level of pathogenic bacteria living freely in the sea, caused by untreated sewage disposal.

Ciguatera

Ciguatera is a very common disease. However, this is not totally accurate: it is not really a disease, but a poisoning that can last for over seven years. The name was given by a captain of a slave ship whose crew became infected in the 17th century when eating the cigua shell, a marine snail. A fitting punishment for those involved in such a trade.

The incidence of human cases varies from 10,000 to 50,000 per year and occurs on a worldwide basis within the tropical regions. It's a food chain toxin, and is produced by bacteria living in algal cells. The host species grows attached to coral structures and is grazed upon by herbivorous fish, which in turn make a meal for larger fish, who then end up on our table. Bon appétit!

But it's not that simple, because at least 3 toxins are involved, though they do not coexist geographically. This results in a zonation of cases, and of the type of fish carrying the toxins. The eastern Pacific has the highest incidence of cases per year, resulting from eating both herbivorous and carnivorous fish, whilst the western Pacific and Caribbean have fewer cases per year and only from ingestion of carnivorous fish. Ciguatera has also been implicated in the demise of the rear monk seal from the coastline of Hawaii.

Ciguatoxin is only associated with carnivorous fish, such as groupers, snappers and moray eels in the western areas, however for residents and holiday visitors of the eastern Pacific, maitotoxin and ciguatoxin are found together in herbivorous fish, such as surgeon fish, wrasse and goat fish, causing serious effects further down the food chain. Scaritoxin has also been isolated, but only from parrot fish. However, when added to the other two causes, it makes an interesting but nasty addition to the many symptoms involved.

Ciguatera carries a particularly nasty set of symptoms. It is not often fatal, but many people who carry the disease wish it was. At first, you will experience severe gastroenteritis, causing you to remain seated on the toilet for long periods of time lasting for at least two days. This is followed by a general weakness between two and seven days into the disease. Beginning on day two and lasting over three weeks, partial paralysis of different areas of the body occurs. This is when the fatalities happen. Death occurs from either severe dehydration from vomiting and diarrhoea, or by paralysis of the diaphragm, resulting in respiratory failure. However, to accompany the above set, other symptoms occur between two and 30 hours after ingestion. These include blurred vision or blindness, burning and tingling sensations in fingers and toes, and the dry ice effect: a temperature reversal where cold seems hot and hot seems cold. Add these on top of the first symptoms, followed by a recurrence of gastroenteritis. This may last for months in the first instance and makes life a real misery.

The toxin is a resilient one, and when the first effects have receded, the toxin becomes deeply seated within our tissues and can remain there quite happily without causing effects for years. But when the infected person ingests uninfected fish, protein or alcohol, they may experience a reoccurrence of blurred vision, the dry ice effect, tingling sensations and partial paralysis. To top this off, if you have eaten a parrot fish which is infected, you will also have a staggering walk and uncontrollable muscle tremors caused by scaritoxin acting on the cerebellum in the brain. The whole set together adds up to making a lot of people's lives a constant misery as, again, there is no antidote.

The disturbing fact is that infected fish cannot be distinguished from non-infected fish. As the areas affected are often very poor and entire villages rely on seafood for their existence either as food or produce to sell, these toxins will carry on infecting many thousands of people per year. A few crude tests can be carried out, and although they are hardly 100% effective, they do provide some protection. The first is to place a silver coin in the guts of a fish, because this area is where the greatest concentration of toxins occurs. If it discolours, don't eat it. The second is to feed the guts to a starved, caged cat (when you are an islander living in poverty, there is no room for animal rights!) and if the cat vomits, don't eat it. To protect tourists, large signs have been erected stating various messages such as 'don't eat the red fish' and 'don't eat the big fish', and on reefs with a long ciguatera history, fishing has been totally banned.

Apart from the horrifying effects this disease has, there is a post script. As mentioned before, biotoxins are important for medical research. It takes 847 kilos of moray eel liver to isolate just 0.35 milligrams of ciguatoxin. Just think how many moray eels it takes to collect 847 kilos of liver—not the whole eel, but just the eel's liver. This type of extraction is very common and wasteful, and out of the world's 108 reef systems only 16 remain untouched. Together with tourism, biomedical extraction, mining and the aquarium trade, to mention just a few, coral reefs will soon be on par with the rate of rainforest destruction.

The Cone Shells

Here we meet our first venom. It's made up of large proteins and, as such, we come into contact with the first poison which we have an antitoxin for. Help is at hand if you are unfortunate enough to be stung, but you only have four hours left at the most if you're lucky!

The cone shells inhabit the tropics and belong to a major grouping with a worldwide distribution. There are about 300 to 500 species of cone shells (the fact that we have to estimate the number of species shows how much we don't know). There are about 10 known species that feed on fish, and there might well be more, but the rest target other invertebrates. It's the fish-eaters that give us humans the most concern, as their venom is targeted against vertebrates. If we think of evolution splitting into two groups, that is, invertebrates and vertebrates, we find different physiological processes specific to each of these two groups. Toxins which target vertebrates will usually have little or no effect against invertebrates and vice versa. So any beast with a venom intended to subdue a vertebrate prey is of concern.

The feeding apparatus of the snail, like that of all snails, is the radula; however, the cone shell has developed a unique radula for the delivery of venom. Instead of having many rows of teeth on a belt to scrape off food from a surface, all the teeth resemble hypodermic needles, and they are loose and kept in a sac, like a quiver of arrows. Venom is passed down a tube and coats a single tooth as it's passed up a proboscis. The tooth is then thrust into the prey. Also, since the tooth is hollow, venom is pushed through the tooth and into the body. In the case of fish-eaters, the prey is paralysed in under one second as a few flicks of a tail would propel the meal away from the slow-moving snail and a passer-by would not hesitate to take advantage of a free morsel (even though it would probably die as well). This form of delivery ensures that no venom is wasted, because it is not energy-efficient to produce and so must not be wasted.

Two main feeding strategies exist: one for nocturnal feeding, and the other for daylight prey capture. The first is where a snail detects a small fish buried under the sand. It extends its mouth and scoops up the fish, stinging it whilst in its mouth. The second strategy involves a brightly coloured proboscis being extended above the sand.

Whilst the snail remains buried, a fish is attracted and then stung in the mouth. In both cases, an extendable stomach engulfs the meal and a happy snail feeds.

But why does it sting man? There is only one reason, and that is defence. These snails have a very thin shell because a lot of the calcium used to construct the shell is used in the venom—an evolutionary trade-off. Better meals, but the cost is less protection. The shell would be of no use against a shell-breaking predator such as a trigger fish, but the venom more than makes up for it. This is reinforced by the fact that the only people stung were handling snails at the time. One case involved a man who decided to place this weird-looking shell in his mouth. Two and a half hours later, he was dead. There have been seven different toxins isolated from this venom, all acting on the nervous system and placing a complete block on all nervous transmissions, which is why it is so fast-acting on fish.

Sea Snakes

These are nasty little beasts if annoyed, but harmless if you leave them alone, and their reputation as being man killers is mostly unfounded. Sea snakes are widespread in the Indo-Pacific regions, but they are absent from the tropical Atlantic. This is because of cold water around the Cape of Good Hope off the southernmost tip of Africa. The cold, up-welling water stops the snakes swimming around and up into the Atlantic. The same occurs around South America. The one passageway that is open to these snakes is the Panama Canal, which connects the tropical Pacific to the Atlantic. Here, freshwater in the central regions stops the spread of the animals.

They have adapted to the marine environment by a flattening of the tail to produce an oar. This flattening gives great water resistance, increasing their propulsion. The snakes also show remarkable oxygen conservation abilities, which has baffled scientists. Trawls have been retrieved after 3 hours, not only are the snakes alive, but they show no adverse effects.

Their venom is a mixture of very potent proteins, all acting on different parts of the body, producing an array of symptoms. The structure of the toxins is not unlike that of cobras and other land snakes, but sea snake toxin, pound for pound, is of a higher potency. They only deliver a very small amount in one go (zero to 31 milligrams per snake), so on a bite-to-bite ratio, many more human deaths occur via land snakes. The reason for this is quite simple: the venom requires a high amount of energy to produce. Sea snakes' prey are often small fish, and as such, they succumb very easily to the venom, so only a small amount is needed to obtain a meal.

The snake's fangs are relatively small compared to land snakes; again, they don't need large fangs to penetrate the body of prey. This has resulted in many frightening, but harmless attacks on divers—the diver is quite safe, as the bite cannot penetrate a wetsuit and thick clothing.

Because of this, it is estimated that only 20% of people bitten by sea snakes show any adverse effects, with 50% of those people dying if no anti-venom is administrated. If the same data was to be collated for cobras, over 90% bitten would die if no treatment was available. There is an anti-venom for each species of snake, but also a universal snake anti-venom if the species is unknown, which is most often the case.

The number of bites that occur each year is very difficult to compile due to the fact that many incidences occur in remote tropical places. Often, those bitten live in villages where transport to the nearest hospital is on foot. Most of these cases go unrecorded, so how many survive or die is not known. We can only collate data from recorded cases.

In one year, 120 bites were recorded, and the human activity at the time of bite is tabled.

Activity	**Cases**
Handling nets	50
Sorting fish	12
Snake trod on	10
Wading to and from boat	18
Washing	19
Swimming	5
Shell collecting	6

When bitten, with enough toxins injected, clinical symptoms include muscular paralysis, pain and difficulty in swallowing and speech. Damage is done to the kidneys and liver and many other tissues, depending on the species. Death is mainly due to paralysis of the diaphragm, resulting in respiratory arrest.

Venomous Fish

There are over 200 species of fish which are armed with a spine and associated venom glands. They occur on a worldwide basis, with the most potent located tropically. Effects range from a simple rash to death; however, an antidote is available for all.
As with all animals that can kill a human, when the press get hold of an incident, it is often reported out of context.

The fish in question only act in self-defence when stinging. They are mainly bottom-dwelling, slow-moving species, and as such the venoms are commonly used as a way of avoiding becoming a meal.

The location of the venomous spines is mainly dorsal or over the operculum plate protecting the gills. They are raised if the animal is disturbed, but they have one weakness. If the attacker holds and swallows the fish head first and in one go, the fins point backwards so the venom is not injected. When being digested the venom is rendered harmless with the digestive juices of a very happy, full predator.

A venom apparatus is common to all species. They have a central spine which is serrated, with barbs pointing backward. This penetrates a body wall easily and lacerates the tissues on retrieval, allowing quick absorption of any venom not already taken up. The barb also has canals running down its surface, acting as channels for toxin flow. Around the barb is a sheath of connective tissue. This houses the venom-filled sacs which are filled by the venom glands, deep within the body. When the barb penetrates a body, the sheath is ripped off and peeled back. This ruptures the venom sacs, releasing the toxins into the canal, which is thrust into the body tissues, allowing absorption of all the toxins.

There are no venom glands actively pumping poison into the body, as often thought. Some species can regenerate the barb sheath after it has been used, but most lose the ability to sting after one use. However, the spine remains (if not initially snapped off)—this alone will act as a defensive weapon.

Injuries related to these fish are very common worldwide. Most people affected are waders and fishermen, both commercial and recreational. The symptoms of a sting often include intense pain, general weakness, nausea, anxiety, vomiting, diarrhoea, sweating, cramps and respiratory distress.

All have an anti-venom, if the patient is taken quickly to hospital. Most deaths occur in remote places, and often this is because of a secondary infection that results from the removal of the barb ripping the tissue. Other deaths have taken place because the patient had a weak heart and died of a related heart attack.

Hydrothermal Vents and Vent Biology

The existence of hydrothermal vents was first hypothesized in 1965 as the oceanic crust's method of cooling in underwater volcanic areas. This idea that the earth literally let off steam at a submarine location gripped the oceanographic community so intensely that the vent systems' apparent existence became the oceanographers' Holy Grail. Technology then was very limited by today's standards, and expeditions were underequipped and underfinanced. A decade later, though, accurate deep-sea sledge towing and submersibles were available for use. One such survey in 1976 used a towed instrument to record hot water spikes and the presence of primordial gases and to photograph giant bivalves over the Galapagos Rift. In March 1977, a series of 24 dives was conducted in the submersible *Alvin* to a depth of 2.5 kilometers on the axis of the Galapagos Rift with the goal of proving the existence of hydrothermal vents: sampling the wine from the Holy Grail.

The Discovery

What happened next can be likened to the realization of Darwin's *Origin of Species* or Watson and Crick's configuration of the DNA molecule. Not only did scientists discover a series of five hydrothermal vents over between one and two kilometers, but dense communities of associated fauna were found living around them. Including red-plumed tube worms up to 3 meters high, giant clams and other bivalves, crabs meandering over the surface, and anemones around the periphery with elongated arms swaying in the currents, a whole new ecosystem had been discovered. Not only had the oceanographers found their grail, but in the process, they had uncovered a new gallery of life. A new field of discovery emerged in the oceanographic disciplines.

Since 1977, a great deal of worldwide expedition has taken place, resulting in the discoveries of new vent fields associated with the 75,000-kilometre-long, mid-ocean volcanic ridge system. In addition these oases of life have now been associated with all areas of submarine tectonic activity: fracture zones, subduction zones and back-arc basins have all joined the rifts in sharing their secrets with science.

The Physical Environment

The questions raised were and are immense: how are these vent systems formed? What is their driving force? How far does their influence spread? How is life maintained there? How are the organisms structured? How are they related to extinct and living animals? The list is endless. And so it will remain, since we are still only at the beginning of our understanding of the vent systems. When we answer one question, there are always two ready to take its place. I feel it is correct to state that hydrothermal vents and the associated fauna are indeed shrouded in mystery. To begin our journey of limited understanding, the geological aspects of these rifts must be examined to determine their influence on the biology of the area, which in turn needs to be examined to attempt to decipher how these remarkable organisms evolved.

Newly-formed basalt rock is riddled with pores and small cracks. Through these fissures, sea water is drawn down to depths of around two kilometers within the undersea crust. This water is superheated to 350°C, and the high pressure there forces it to return to the surface. Along the passageways, the water collects a mixture of chemicals found there, the specific compounds depending on the individual site. This potent soup, referred to as hydrothermal fluid, returns to the surface through the fissures and is exhaled up into the water column where it mixes with the surrounding waters.

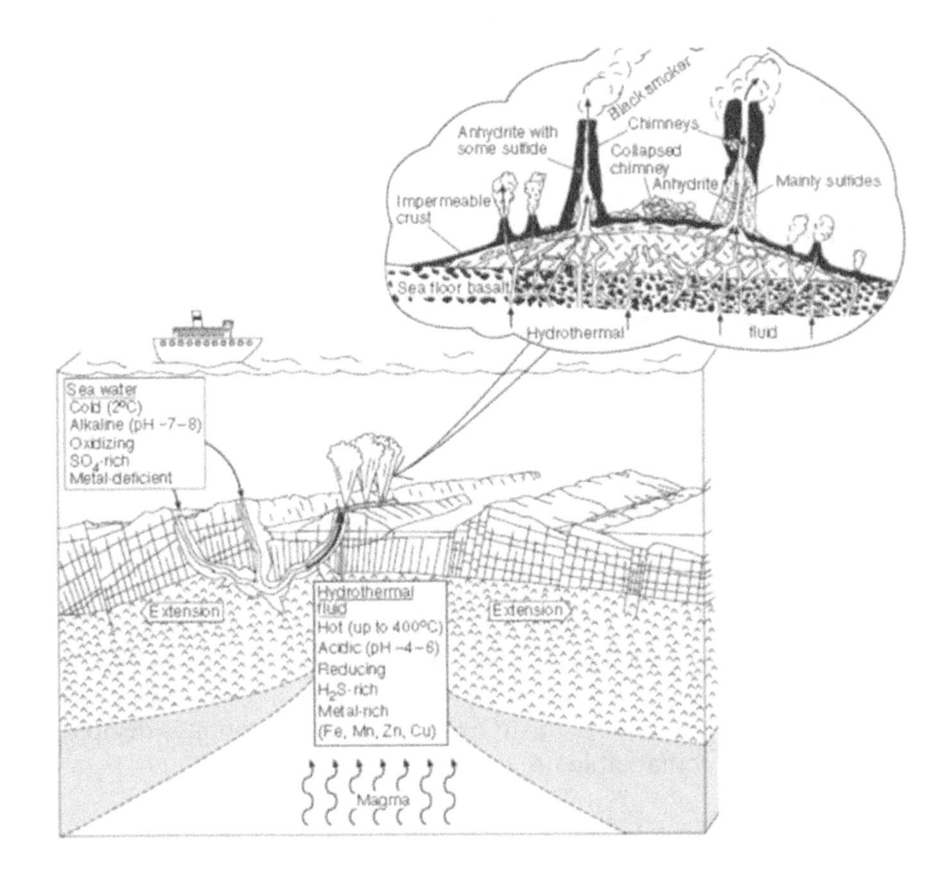

Passive venting where the fluid returns at a low velocity through cracks and fissures at 5 – 17°C. The temperature is decreased because the returning fluid is greatly diluted by sea water while it is still within the basalt rock. This subsurface mixing reduces the velocity of the upcoming water and results in shimmering water being extruded from the cracks.

Fluid flows violently at 250 – 380°C through undersea chimneys known as black or white smokers in a process called venting. White smokers experience a small degree of subsurface mixing, but fluid remains pressurized and is forced to the surface. This fluid appears a whitish color due to the dilution of chemicals via mixing. Black smokers are the most powerful, with fluid reaching temperatures of 350 – 380 °C. Mixing occurs between the subsurface chemicals and sea water, and these large sulphide chimneys are built up rapidly. Fluid mixes with the deep waters, which have an ambient temperature of 2°C, to produce a localized temperature of around 14°C. From this a precipitation of chemicals occurs, building up the chimney and coating the surrounding rocks on which the fauna exists. This process results in an environment of locally raised temperatures with high concentrations of sulphur compounds either in solution or precipitated around a raised chimney.

This localized habitat, when colonized, can be classed as an ecosystem, because the surrounding waters form a physical barrier within the constraints of temperature, water flow and water constituents.

Since the discoveries, paleontologists have positively identified six fossil vent communities with possibly a seventh dating from 450 million years ago. These sites have revealed a common linkage through time and space in relationships existing between extinct and living fauna in terms of their community structure. The same groupings of animals existed around ancient vents, expressing the same low diversity, abundance and unequal distribution of species as present vents. Ancient vent systems were just as isolated from the surrounding waters and inhabiting species as modern vents are today. This suggests that the endemic fauna has existed in a state of genetic isolation for hundreds of millions of years, resulting in restricted gene flow within the vent community or between neighboring vent communities. Remarkably, these communities have passed successfully through geological time, apparently escaping the mass extinctions that occurred within the marine environment, probably due to their isolation and dynamic nature. The current thinking is that a high degree of large mineral deposits which are actively mined today are the result of ancient hydrothermal vents, due to the localized precipitation.

Vent Biology

Success in a biological system has to be based on acquisition of energy: this is the basis of every organism's life history. Many examples exist of behavior evolving to maximize the amount of nutrition available to the species or individual. Salmon undergo great migrations through the oceans, animals lie in ambush, toxins have been developed to subdue prey; the list is endless. However, all depend on plants for primary production to support the food chain. Yet in this world of constant darkness there are no algae on which animals can graze. Instead, life is based on the ability of bacteria to oxidize sulphur compounds from the hydrothermal fluid, thus generating energy for life. Similar examples exist in anoxic conditions in a variety of habitats throughout the world, but none have evolved to support a whole ecosystem of multicellular organisms. This raises an important question: were these areas the first to support animals from the primordial soup, since the atmosphere was also oxygen deficient then?

Modes of nutrition in the larger animals, as expected, range from simple grazing and predation to the most complex symbiosis. Their efficiency is unrivalled. To date, 375 species of animals have been recorded living at vent locations, of which 93% are new to science and so have no common names. I will not bore you with the scientific name, so we will refer to them by more generic names.

Anemones and hydroids are associated with the periphery of the vent. Due to the difficulty of retrieving individuals, no species has been positively identified. They are collectively known as siphonophores. Situated under overhangs or erect on the upper surface of the substratum, the tentacles reaching up to 1.5 meters in length are orientated into the currents plying for zooplankton or organic material brought into the area from the surrounding waters. Densities can be vast, indicating fission reproduction (in which the organism simply splits in two) as well as sexual reproduction, to form carpets of tentacles reaching up or down into the waters.

Two main species of mollusks, clams and mussels, dominate. Some truly are giants, reaching 20 centimeters across. Because they are easy to retrieve, they have been some of the most studied organisms of the vents. Both species contain symbiotic bacteria, which converts sulphide compounds to energy, existing at different densities in their gills.

These bacteria produce most of their nutrients. Mollusks can survive away from the vent by filter feeding, but they reach the highest densities near the vent outlet.

Giant clams are more restricted to the crevices where they experience a slight elevation in temperature. They insert a large foot into the water flow through these pathways, absorbing sulphides and transporting the compounds to their symbiotic bacteria in the gills and those in the foot. Water flow has a profound effect on the community, and any alteration results in either repositioning to adjust to the changing flow or death. Sexually mature clams have been seen at ages between one and four years, though fully developed sex organs are only apparent within a size class of 12 – 14 centimeters, suggesting that maturity is dependent on size instead of age.

The mussels have a reduced gut and mouth that is thought to actively filter feed as well as receive nutrients from its symbionts, though this still remains under-researched. Male mussels mature at two to three years while females are fully mature at four years. Hermaphrodites exist in the intermediate stages, suggesting that some of the population change sex with age.

Limpets are the most diverse of the hydrothermal vent fauna, with 30 species recorded to date. Their close relatives are from the Paleozoic, but at least one archaeogastropod still in existence is from the late Mesozoic, that is 240 million years ago, making it one of the oldest surviving species to date. These are found in various places in different zones and sites, and aside from one filter feeder, all limpets graze over the hard substratum and even on the bivalve shells.

The giant tube worm is the organism that has advertised the hydrothermal vent fauna to the general public. Nine species have been recorded to date, with tube sizes ranging from 10 centimeters to three meters, though only one species is the giant. They are highly dependent on the vent fluid flow and so form dense clusters around the base and on the chimney which provide a habitat for other organisms. Their external morphology is simple: their only movement is the retraction and extension of the red plume; they have no mouth or gut, instead absorbing sulphur through the plume and like the bivalves, they support a very high density of symbionts to provide all their metabolic needs.

Crabs are abundant in the hydrothermal vent's ecosystem. They obtain nutrients by scavenging and grazing and have been observed touching the tube worms' plumes without retraction occurring. Do these crabs perform a cleaning duty like various species in the shallow seas do?

Shrimp swarm in vast numbers within and around the venting fluid. Little is known about this animal, but one of its features is currently the concern of many biologists. Like no other animal we know of, the shrimp's eyes are located on the bottom side. They may be the only natural visual heat detecting apparatus known, apparently used to observe the heat flux of the venting fluids. This may have evolved as a self-preservation feature, since the shrimp have to remain near the venting fluids but must avoid getting too close to this dangerously hot environment.

The animal groupings discussed above are all endemic to and typical of the vent ecosystem in their distribution, but this is not to say that individual species inhabit each vent. Major distinctions occur between species at individual vents within the same field, and species divergence between vent fields on a global scale has progressed, though a general mix of hydrothermal fauna exists. Some vents will exhibit an exclusion of species, reducing the species diversity. One example is a vent in the Mid-Atlantic Ridge that exhales no hydrogen sulphide. This results in the total dominance of siphonophores, which are normally restricted to the periphery of a vent. Other vents represent a similarity in the community structure related to nearby or geographically distant vents. An excellent example of this is the discovery of a venting field in the North Fiji Basin in 1987 that showed species of typical vent animals previously discovered in the Galapagos Rift region in 1977.

The community structure has been shown to be very dynamic and hostile to the inhabiting organisms, but an even greater hostility can be seen in the physical forces driving the ecosystem. First, the vents are on top of an active volcanic area subject to tectonic movement. Second, inhabiting fauna all depend on the diffusion of sulphur compounds into the surrounding waters, so much so that it has been estimated that 75% of species are dependent on symbiotic bacteria. These two features exert such a dominant force on the biology of the area that competition becomes a community-structuring force only when the geological activity allows it to proceed.

In 1985, during expeditions to the Rose Garden hydrothermal vent located within the Galapagos Rift, comparisons to the animal community structure were made to the structure observed during previous visits in 1979. Significant changes in all aspects of the structure had occurred in just six years. To the observers' astonishment, the whole community structure had shifted. Where tube worms had dominated, they now were in very low numbers, and bivalves flourished around the chimney. Siphonophores were totally absent, whilst mobile scavengers and predators were more abundant. Every species had shifted its position in the community.

The partial collapse of the vent chimneys had caused this mass dislodgement of the tube worms. The redirection of the vent flow was the major cause of the shift, effectively removing the food source for the chemosymbiotic bacteria within the worms. A lack of food caused their death, allowing the more efficient bivalves to move in and take over. Eventually, in an event like this, the flow shift is so profound that the venting ceases completely, causing a localized extinction. When this is happening, though, a new vent is born elsewhere in the field.

A general life cycle of the vents has formed: The tube worms will colonize and dominate the early stages, then bivalves will appear, with the mussels dominating as the tube worms decline. Siphonophores and other suspension feeders will colonize the periphery and surrounding substratum, but these populations also decline as the venting fluid slows. Then mobile scavengers and predator populations will increase feeding on the dead and dying animals, and move to surrounding sites when the venting ceases. This pattern will vary from site to site depending on the fauna.

The estimated life span of a vent is variable; the time from birth to extinction is between 30 and 100 years. It is important to put this into perspective: a complete ecosystem is born and colonized, biological succession occurs, and the life there becomes locally extinct within 100 years. Other ecosystems all over the earth have been developing since life began. This fact alone puts immense selective pressure on the organisms to mature quickly and reproduce with an efficient dispersal strategy to allow the young to find a new home and start the process over again.

We have examined the different types of animals inhabiting vent systems, the differences in the type of venting that occurs and the short life cycle of vent ecosystems. So what animals inhabit the vents that have been explored? The picture of a vent that most people imagine is one of huge black smokers surrounded by giant tube worms; however, this cannot be further from the truth. Of the areas explored to date, distinct species differences occur across various geographical locations. The Galapagos rift has the only known populations of giant tube worms, and anemones that are found only here. Along with clams and mussels, this area has the highest number of different species of any vent field. The East Pacific Rise and Juan de Fuca Ridge Systems have small tube worms, very few clams and no mussels, but they are home to many other species not found at the previous site, though the East Pacific Rise has nearly twice the amount of species than the Juan de Fuca. The Mid Atlantic Ridge shows complete separation with totally different species of crabs, mussels and worms and huge swarms of shrimp up to 2500 cubic meters in volume. To discover why this is, we have to travel back in time 65 million years and examine the ridge systems themselves.

First we will look at the ridges themselves, as we are situated on a globe and not a flat surface. The ridges are broken up into many segments, each between 30 and 90 km long. Some segments have no venting, others have many vent fields and some have only a couple. Within 50 years the whole situation will have changed: a previously inactive segment could now be riddled with vents and others could become completely extinct. Each segment is separated by what is known as a transform fault, across which flows a different water current. Think of a balloon (the Earth) then put a wooden ruler on it, bend the ruler until it breaks. The ridge has now broken with the shape of the earth, the break in the ruler is the transform fault. So when the larvae of the populations of one segment try to cross this fault, many if not all are lost, isolating the population of each segment. This separation means evolution occurs at its own rate on each segment. However when the water flow across the barrier allows them to, some larvae will make it across. They then find a vent and settle, assuming that vents are active on their new segment, allowing some genetic distribution between different populations to occur.

Consider our motorway systems. The roadways are long, but there are only a few isolated areas separated by distance where you can get refreshments. These service centers are like vents; there are plenty of areas along these currents, but only a few isolated areas, with different distances between them, where larvae can settle. If vent location was the only factor, then all vents would have similar animals living on them. But some of our motorways are experiencing road works meaning some destinations have been cut off and are now unreachable. Cut-off currents result in no input or export of genetic information for the next generation. Some new roads have been built to connect the motorways. New currents allow genetic information to be passed between sites previously unreachable. These undersea roadworks have been in progress since time began. The result of such isolation or connection is either a complete separation of species or the persistence of genetically static conditions. One such example is a sessile barnacle that dominates one area in the Mid-Atlantic. This is the most ancient known species, dating from the Mesozoic era and found nowhere else.

So let's travel back 65 million years and look at the main force behind such occurrences. At this time the Mid-Atlantic Ridge system was not connected to the Pacific, with the north and south areas of the Atlantic only just opening. At 20 million years ago, we find that the ridge system is connected to the Pacific and Indian Ridges via the Antarctic Ridges. However, large transform faults bar most larva dispersal. All animals here are totally separated from the Pacific fauna, resulting in what we see today as a completely isolated community. In evolutionary terms 20 million years is a short time, so even if there was a good connection only small differences would be evident.

In the Pacific 65 million years ago all the ridges were connected, and communication was possible along the whole length. If we look at the species number, we find that the highest diversity is located along the Galapagos Rift, so we assume that species moved away from this area to colonize the Pacific System. At the junction of the two ridges, we have a three-way split that proves a major barrier and results in species dilution along the Pacific. 37 million years ago, tectonic plate activity split the Pacific ridge in two by the continental land mass of the United States. The result of that split is now known as the San Andreas Fault. This divided the vent populations in two: first, there was a higher diversity along the East Pacific Rise, and second, farther north, there was a low species number due to isolation and more transform faults across which the larvae could migrate. That is why there are completely different communities of animals in different vent systems and why some species, like the giant tube worm, are only found at one site.

We have only known about vents for 38 years, and the fact remains that the majority of ridge systems have not been explored. We don't know what inhabits the Indian Ocean Ridge or the Antarctic Ridge system, among many other smaller areas. Our knowledge is limited by our funding, as the cost of running a research ship can be £25,000 per day. Until these areas are studied, we only have a small portion of the whole story. We must admit that what we know about these systems is only based on our current understanding; the majority of what we think we understand might well need revising.

But to throw a spoke in the wheels, recent discoveries of clams, mussels and worms that are related to vent communities have been found in the abyssal zones in the mid Pacific, thousands of kilometers from any vent system. These animals when in a larva state recognize the unique chemical state of the vent habitat that allows them to settle; otherwise, they would die out. So what conditions allows these animals to live here? The decaying body of a whale will produce the same sulphide required for these animals to survive. Somehow a lucky larva from a vent site was carried over a dead whale, where it settled. Who knows where this situation will lead when we think about the unexplored areas and what might be lurking there.

There is one thing certain: in the depths of the oceans, in a totally anoxic, extremely dynamic environment exists a natural phenomenon of unrivalled complexity in nature. Science has just scratched the surface of understanding and unravelling its many mysteries. It is, beyond any doubt, the most important ecological discovery of the 20th century, and if in my lifetime there are tourist trips down to visit them, I will be the first to buy a ticket.

Life in the Polar Seas

When we think of the polar seas, we always envisage a very hostile environment mainly due to two factors. The first factor is the cold temperatures that freeze hundreds of square kilometers of sea surface every year, and the second is the rough weather and often extreme storms. Because of these, we always think that life here must be very hard for the inhabitants—but how wrong we are. In this situation, we are taking the human viewpoint—and indeed, life for human populations in polar regions is very hard because we did not evolve here and are not adapted to the harsh environment. Life for the local inhabitants, however, is no harder than it is for a sturgeon fish to swim around in a coral reef. If we took that sturgeon and placed it in a polar sea, it would die in a short time from cold. But if we took a polar cod and placed it in water at a temperature of 8°C, it too would die, this time from heat shock. All animals, both warm-blooded and cold-blooded, have evolved to make this habitat their own. We will look at how these cold temperatures and the adaptations of the animals have resulted in a unique ecosystem.

However, this has not always been the case with cold temperatures in the Antarctic. Between 500 and 600 million years ago, in the Cambrian period, the tropical regions of the earth were a little warmer than today. However, both poles were covered in ice, with deep water constantly at 2°C. In the Triassic and Jurassic periods, 200-300 million years ago, the whole globe was tropical and had warm deep water. This resulted in a mass extinction of cold water species and a global rise in sea level.

Around 60 million years ago, the Antarctic showed another shift in temperatures, finally reaching the present day situation. Since then, Antarctica has had an ice-covered landmass and deep water temperatures of 2°C that are consistent worldwide. We have had stable temperatures for around 60 million years. This stability, alongside the fact that the separation of the Antarctic landmass occurred very early in geological time, results in a high number of species that occur here and nowhere else on earth.

The Polar Environment

Let's first look at the temperatures experienced in the Antarctic and how its inhabitants have adapted to them. Inland areas experience fluctuations between -10 and -60° C, with the wind chill factor making it feel much colder, but inland environments are rather irrelevant to sea life. The stretch of land that meets the sea in the intertidal zone experiences temperatures between -5 and -10°C, but with ice scraping the rock surface constantly, very little life exists here. This proves a stark contrast to other intertidal areas on the globe, which are extremely abundant in life.

The sea temperature in the summer hardly ever rises above 0°C, but the salt content ensures that the sea remains fluid instead of freezing like pure water would. In winter, it's a different situation, the surface temperature dropping to -1.9°C—the freezing point of sea water. This results in many hundreds of square kilometers of sea surface covered in ice. The animals that live here must overcome this hostile environment to survive.

The endotherms, that is, warm-blooded animals, have to remain at their core temperature to function; when a rise or drop in temperature is experienced they start to suffer and cannot compete with others. If this temperature deviation is serious, then vital proteins such as hemoglobin, which carries oxygen in the blood, cannot function, so the animal quickly dies. This is why hyperthermia and fever temperatures are so dangerous. In the polar sea, we have two types of endotherms: the mammals which have a core temperature of 35 – 38°C and the birds living at 40 – 42°C, the differences caused by their different paths up the evolutionary tree. We have a large discrepancy between the animals' internal temperature and that of the environment; therefore, to maintain the working animal, two things are needed. The first is insulation, a need supplied by both behavioral and physiological adaptations, and the second is a good food supply for heat that is maintained via the heat loss from all the chemical reactions that occur within the body. A low food supply plus a low temperature equals death from cold before starvation.

Animal Adaptations

The one thing everyone knows about Antarctic mammals is that they are fat: they carry 30% more blubber than temperate relatives, but they have a unique adaptation to this blubber. Fat conducts heat at a very low rate, thus body heat is retained, but only if warm blood is not passed through it. So the animals have a problem: they have to supply skin cells with blood to keep the cells alive, but to do this constantly would result in vital heat being lost and the individual swimming in the heavenly seas. To avoid this untimely demise, the capillaries have valves on them, allowing blood through only when needed, thus keeping heat loss at a minimum. In addition to the blubber, many animals are covered in fur, which traps warm air next to the skin. This is a very good adaptation when lazing around on the beach but its effectiveness is low when in water as the trapped air soon cools down and escapes, so fur works when resting and blubber comes into its own when hunting.

Another famous behavioral adaptation in the Antarctic is clustering: great numbers of animals gathering together to find shelter and shelter one another. Clustering has its costs, though: it often results in fighting, which wastes vital energy, and ice breaking, which results in the animals taking a dip. The Arctic walrus shows the greatest parental care of all the polar animals in regards to keeping warm. When a baby walrus is born, it's naked; wet with no fur and little blubber. It would be dead within minutes if the mother did not shelter its offspring and give it body contact. Where the baby is touching the mother, the blood capillaries in the blubber not only open but actually enlarge, allowing a greater conduction of heat from the mother to the baby, warming it and keeping it alive. The small one soon produces an effective layer of fat due to the very high fatty content in the mother's milk. Birds here have little problem keeping out the cold when above water with their feathers, but feathers are only effective for short periods of time when underwater. In polar regions, the birds have oilier feathers and spend much more time preening (waterproofing) than their temperate relatives. This is also the reason why penguins have lost theirs in favour of fat.

The Fish

So we come to the poikilotherms, that is, the cold blooded fish and invertebrates: these animals function at the same temperature as their external environment. Because life for cold-blooded animals is very slow, many fish reproduce after seven years, whilst their temperate relatives spawn at one year old. That's a bit misleading, because fish worldwide reproduce after reaching a mature size, not age, so in the polar sea we have a fish taking seven years to grow to the size of a one-year-old in warmer waters. Many invertebrates take 20 years to reproduce, and some individuals have been found to be over 300 years old. The oldest living animal ever discovered is an Artic bivalve. Every year they lay down a new layer to their shell, like trees. So if we count the rings, we can tell the age of the creature. The oldest one was in its larval stage and settling into its adult habitat when Queen Elizabeth 1 was on the throne in England. The animals here also attain giant sizes, one example being an isopod (think of it as a marine woodlouse) that grows to be over 30 centimeters long, now that's one big woodlouse. But how do these animals function and remain very successful at these low temperatures, and what dangers are there? Here we will split the two groups and deal with the fish and invertebrates separately.

The polar fish live a more dangerous life than the invertebrates because of their freshwater origin. Their body fluids have a lower salt content than the sea around them and a fish will start to freeze between -0.5 and -1°C when the sea water freezes at -1.9°C. There are two different ways in which freezing injury can occur. The first is when ice forms in the cells and the salt content of the cell increases to toxic levels, so if the ice doesn't kill, the salts will. The second is with the formation of ice itself. We have all seen ice crystals on cars and observed the elaborate shapes they form; however, in biological systems, ice forms differently. Here instead of nice flakes forming, a single spear grows, very rapidly. This ruptures the cells and kills them, and a fish can be dead within seconds. So the first problem is that the water will remain fluid at a temperature that will freeze the fish. However it's not as simple as that, because in ice-covered waters, tiny ice crystals exist down to a depth of 30 meters, where the water temperature is just above the fish's lowest limit. So we have a fish swimming happily, without a care in the world apart from being eaten, and suddenly it swims into an ice crystal. On contact, the crystal acts as an ice seed, causing the formation of ice throughout out the fish, and bang! Our fish is dead.

Two ways of overcoming this hideous death have evolved. The first is where the species perform short migrations to the deeper, warmer waters, thus avoiding death. These species' young find life a little more risky, since they are too small to undergo such travels, so every winter they bury themselves in the soft mud, taking refuge in the thermal insulation found there. But if an ice crystal touches them, their mud refuge becomes their grave. The second adaptation is most remarkable, and allows the fish to swim through ice-laden waters and perform naturally. I have stated that a normal fish's body fluids would freeze at -1°C; however, these fish's bodies freeze at -2.2°C. This separation of freezing points is due to a biological antifreeze. It works in the same way as the antifreeze you put in your car every winter. The only difference is that it is a lot more effective than the stuff we manufacture. In a car we have to fill at least one third of the cooling system with antifreeze, but only 3-4% of an Antarctic cod's are antifreeze. We can manufacture the molecules synthetically, but it's too expensive. These two adaptations have allowed many species of fish to inhabit waters that would be deadly to any others.

The Invertebrates

For the invertebrates, life is a bit less complicated. They started life in the sea hundreds of millions of years ago, so their body fluids have the same salt content as the sea, and will only freeze if caught in the surface ice-forming layer. To avoid this, they just move to deeper water, or go down in the sediment. However, some species, such as the mussel, can be frozen in sheets of ice and still function normally when thawed. They achieve this by expelling all unwanted water from their body, increasing the salt content of the cells and lowering the freezing point of the body tissues. This is so effective that mussels have been frozen to -10°C and winkles down to -22°C, then thawed and returned to a normal active lifestyle.

So we have seen how life adapts to living in such an environment, but what life do we find in the polar seas and why is the food chain so different there than anywhere else on earth? There are two main reasons why life and life processors in the polar seas are so unique to the area. The first is that daylight only occurs for 120 days per year but during that time, it's for 24 hours a day. The second is due to the temperature of the sea; the colder the sea water is, the denser it becomes. The old theories are that the polar seas are extremely productive and support an immense array of life in high concentrations; this is now known to be untrue. When the first sampling expeditions were carried out, it was done in the summer, and in a coastal area, at a time when the whole ecosystem is poised for growth, so a full year's sampling was not carried out. We now know that life here is no richer than in the temperate seas, but it certainly behaves in a different way.

Therefore we come to the basis of all food chains: the plants, or in our case, the single-celled algae. The one thing that they have in their favour is that the waters are full of nutrients to allow rapid growth. So in the summer, we would expect to see a body of water full of algae, but the prevailing wind of the Southern Ocean is never less than force 7 (about 50 – 60 kilometers per hour), so as soon as an algal cell grows it's carried down 600m by the mixing effects, feeding the bottom waters. In the sheltered inland areas and underneath the receding ice sheets where there is no mixing, extremely high concentrations of algae occur. These are further increased by the algae seeds being released by the ice melt waters, where the water can turn green due to the high concentration.

Another factor comes into consideration here: the cold, dense waters. Water of this density can support single cells up to 150 micrometres long, truly a giant by algal cell standards. If we compare this to temperate and tropical algal cells of 20-40 micrometers and 10 micrometers respectively, we can see the difference; algae size has a very significant effect as we pass up the food chain. So in the offshore areas, algal growth is rapid, but is lost just as quickly. The inland seas experience a very high population, but all growth and production occurs in the 120 days of summer. I mentioned that life is slow due to the cold. Another reason for this is the summer days, when all animals go into overdrive feeding to build up reserves, for a meal is very hard to find in the long drawn out winter months when most feed off the built-up body reserves.

Here we move up one place in the food chain into the realm of the herbivorous planktonic animals. Plankton means drifting life; it does not mean small. If an animal cannot swim against the current it is a member of the plankton; even large jellyfish are planktonic organisms. When I say herbivorous planktonic animals I do not mean small ones; I mean the Antarctic krill that is 6-8 cm in length. Over 90% of animals caught in plankton trawls are *Euphausia superba* the Antarctic krill, which is an amazing animal existing in huge numbers, the estimated wild stock being around 200 – 600 million tons, give or take a million, which is quite a few shrimp when you think about it.

These krill live in huge swarms a few hundred meters below the sea surface during the day, but when night approaches, these swarms rise to the surface and disperse feeding, only to regroup and sink when fed. This is their one big mistake, for when grouped they make an easy meal for the whales, or show up easily on echo-sounders in trawlers. When a whole group is caught, the few survivors find other groups, only to be taken later. This could be their real downfall as the fishing industry is increasing. They are vulnerable not only due to their social behavior, but also due to their life history. Being an Antarctic species, their life cycle is slow: they only reach sexual maturity at around five years, and they live to be 25 years. We can see the difference when we consider that its tropical relative is able to reproduce after 14 days. It takes a long time to replace a stock that is taken and, as with all Antarctic species, their future is in question due to the pressure of increased fishing.

So what feeds on the krill? A better way to pose this question is, what doesn't? The shortest food chain in the world is this: Antarctic phytoplankton cells are eaten by krill, which are eaten by whales. The cold, dense waters make this possible. If we had the same size krill in temperate waters, the animal would not be able to feed on the algae, as it could not handle such small cells, so the cold water and its properties are very important in allowing relatively large plants and animals to exist. Many others make the humble krill their staple diet: most of the hundred species of fish living in the Antarctic, the 10 million tons of squid, which eat around 15 – 20% of their body weight per day, more than 40 species of birds, the crabeater seal, the fur seal and the great whales. When you think of the Antarctic food chain, think of krill and imagine what would happen if they were overfished: fewer krill – fewer squid – fewer sperm whales – more squid – even fewer krill – fewer baleen whales. A simple food chain like the one in the Antarctic can easily be disturbed!

So what about the rest of the inhabitants? Let's see what covers the sea bed and makes use of the dead algal cells that fall out of suspension due to the stormy weather. There is so much of it that it's called marine snow due to the sheer number of cells and their whitish color. In the shallower waters, around 10 meters deep, life is very common: feather worms catching the food in their peacock-feathered plumes, sponges, bivalve shellfish filter feeding and starfish eating the shellfish. All grow to giant sizes and have very long lives if they are allowed to reach old age. If we move down below 100 meters deep, the scene is quite different. Sponges make up 74% of the total fauna, living in an amazing density. There is a total of 134 grams of dry weight per square meter of sea floor (the dry weight is the weight of the animals when all the water is removed). When we compare this to just 55 grams of dry weight per square meter in temperate waters, we can see just how important the huge amount of marine snow that falls from the surface waters is in driving deep-sea life.

The fish living here are nearly all related; most belong to the polar cod family, but in total there around 100 species known to science. Very little is known about their lifestyles. What is known is that they remain inshore for their first 4 years, not growing much, but just existing. After this, in the spring, a mass migration to deeper waters occurs where they feed actively on krill. Next, they grow rapidly, putting on reserves for their first winter offshore. They breed at between 7-8 years and can live to well over 50 years. Increasing fishing pressures are being experienced due to the fishes' increasing scarcity; the 400,000-ton wild stock that has been estimated for one species won't last long.

The Birds

When naturalists first visited this area, they came across islands to the north of the polar front. Here, they catalogued vast numbers of primary sea birds. This caused the early scientists to think that this area was the richest in sea bird species and numbers, but again, they were wrong. The huge aggregations that were seen was due to two factors. First, they have no land predators to kill them or take eggs. Second, there are very few sites on this continent that are suitable for a breeding due to adverse winds, so they form huge aggregations. There are a total of 50 breeding species of sea birds here, a number comparable to the temperate zones. However, when calculated, there are only 0.12 milligrams of bird biomass per square meter of sea surface (oh, how scientists measure things), which is much lower than in temperate zones. Most species feed on krill and can delay breeding to be synchronized to the krill increase when algal blooms occur. They also take fish and squid, and pick at any floating dead animal. One species of albatross targets sperm whale sick trails, in which it finds some tasty nibbles: mainly partially-digested squid. Bon appétit! Many species of birds co-exist and feed on the same food source, yet competition here is non-existent due to behavioral traits. The Macaroni and Gentoo penguins live at the same sites in huge numbers, and they have to feed and feed their young on krill at the same time. The Macaroni penguin swims far offshore to hunt and takes a 35-hour feeding trip, whilst the Gentoo penguin feeds inshore, taking only 9 hours to complete a trip, so the Macaroni effectively swims over the Gentoo's food supply. This is how so many species of birds can coexist whilst feeding mainly on the same food source. Populations are doing very well; they doubled between 1940 and 1960 and they did it again from 1960 to 1980. The Antarctic is indeed a haven for birds.

The one thing that sticks out like a sore thumb when discussing the life here is the huge densities that they exist in, and also the very low diversity of species found here. To answer this, we have to look at the environment, for life has existed here for long enough to produce a wide variety of species at least somewhat comparable to the temperate regions. What we find is that there are very few ecological niches within the ecosystem: there are no marshlands or swamps, intertidal life is virtually non-existent due to ice, there are no kelp forests or sea grass beds and no coral reefs; if it's not bare rock, soft sediment or ice, it doesn't exist in the Antarctic. There is very little variation in the available habitats to allow species to evolve and colonize different areas. Another important factor lies in the long generation time: it takes a long time to reproduce, and when they do, it's in much smaller numbers than in other regions of the earth. The whole secret to evolution and the emergence of new species lies in the replication of

DNA. To put it simply, when a mistake or mutation occurs in the new DNA, this represents the first step in the evolution of a new species, but it takes time; in fact, it's between 1 and 20 million years for a new marine species to evolve. So when organisms reproduce at a rapid rate, evolution is faster than those with a slow reproductive rate. A barnacle in the tropics spawning every 14 days will evolve into a new species a lot more quickly than its polar relative that reproduces once per year. However, when we add these two factors together, we also have to look at another factor: the Antarctic covers only 6% of the world's surface. The area which has the greatest richness in species is the tropics, where we have fast reproduction and lots of different niches in the ecosystem—and lo and behold, it covers 43% of the earth's surface!

The Mammals

The only animals that we haven't mentioned are the large mammals that exist in the polar seas, and to be honest, there's not a lot to be said that isn't covered in great detail elsewhere due to their charismatic species status. However, a few points are not coved in great detail. One is the ability observed in some whales to reproduce at an earlier age when populations fall, the Minke whale used to reach maturity at 14 years of age, but now they reproduce at 6 years and the population has doubled. The question that has to be raised is, will their reproduction age return to normal, or is this is now the current biological state? We don't know. Some whale stocks are now showing a small recovery due to their long generation time. Most breed every 2 years and produce only one offspring when they reproduce, so it will be a long time before we know if they will make a full recovery. For the Blue Whale it might well be too late and they could be caught in an extinction vortex where their numbers are too small to allow an increase. The single biggest threat to all whales visiting the polar seas is the threat to the animals lower on the food chain like krill. To over-exploit these will cause a collapse in the whales' numbers, and the possible extinction of some species. This is a real possibility; many species of polar fish are already under threat because the remoteness of the area makes governments unwilling to spend money on fishery policing. Soon, the krill will be targeted by large fishing fleets capable of removing thousands of tons at a time. The krill will be under threat, and I can't imagine a worldwide outcry on the demise of a shrimp. It's a pity, because dolphins and tuna are not the only animal worth saving. When this happens, and it will, then all the worldwide programs to save the whale and all the money that was raised by the supporters' hard work will have been for nothing.

To save the whale, we have to save the krill. It's quite a sad note to end on, but maybe a higher profile in the roles such animals play in the game of life should be promoted to ensure that this does not happen.

Index

A

B

C

D

E

F

G

H

I

J

K

L

M

N

O

P

R

S

T

U

V

W

Z

Marine Life
A three-book series by
Andrew Caine

Marine Ecology for the Non-Ecologist

ECOLOGY – THE BASIC FACTS

THE PHYSICAL ENVIRONMENT
THE BIOLOGICAL ENVIRONMENT
FOOD CHAINS, WEBS AND ENERGY FLOW
RANDOM 1 – THE IMMORTAL JELLYFISH

THE PHYSICAL ASPECTS THAT SHAPE THE COASTAL ENVIRONMENT

WATER MOVEMENT
WAVES, TIDES AND CURRENTS
TEMPERATURE
SALINITY
RANDOM 2 – THE BLUE DRAGON

THE ROCKY SHORE

THE MUD AND THE SAND

WHERE RIVERS MEET THE SEA

THE MARSHES AND THE MANGROVES

CORAL REEFS

INDEX

Incredible Oceans!

Amazing facts and explanations from the wonderful worlds of:

Marine Biology
Marine Ecology
and
Oceanography

THE OCEAN

IT'S ONE BODY OF WATER AROUND
THE GLOBE AND IT'S AMAZING.

THE PLANKTON

YOU CAN'T SEE MOST OF THEM.
JUST WAIT UNTIL YOU HEAR ABOUT THEM; LIFE-GIVING
YET DEADLY AND SILENT!

LOCOMOTION AND MIGRATIONS

FROM THE SLOW TO THE SUPER SPEED,
IT'S ALL A MATTER OF LIFE AND DEATH
GOING ON HOLIDAY AND WHY! NO
WAITING IN AN AIRPORT AND NO
SAT NAV EITHER.

FEEDING

FINE DINING TO FLUID ONLY,
NO TABLE MANNERS HERE

REPRODUCTION AND LIFE SPANS

PASSING ON THAT DNA, OR JUST MAKING
A COPY, MIND-BLOWING ACTIVITIES AND
STRATEGIES.
LIFE FROM HOURS TO 100S OF YEARS, WHO
WAS SWIMMING WHEN QUEEN ELIZABETH 1
WAS ON THE THRONE.

HOUSING

ONE OF THE BIGGEST SHORTAGES TO HIT
THE OCEAN LIFE, REAL ESTATE. WE ANSWER
THE IMMORTAL QUESTION –
WHO LIVES IN A HOUSE LIKE THIS?

RELATIONSHIPS BETWEEN SPECIES

WHO HELPS WHO AND WHY, OR PARASITES?
TO SUCK YOU DRY?

WHATS GOING TO KILL YOU?

YOU DON'T WANT TO MEET THESE - EATING ONE IS DEADLY TOO.

POLLUTION AND DESTRUCTION

THE SOLUTION TO POLLUTION IS DILUTION. A SAYING ORIGINATING FROM THE 1960s. ONLY NOW THE OCEAN IS FULL UP!

Made in United States
Orlando, FL
15 May 2023

33140547R10093